635.9774938

LONDON BOROUGH OF LEWISHAM

LIBRARY SERVICE

Author

Title

Bamboos

A. japonica, Botanic Gardens, Dublin. By courtesy of the Irish Tourist Board

BAMBOOS

A Gardener's Guide to their Cultivation in Temperate Climates

Alexander High Lawson

FABER AND FABER LIMITED
London: 24 Russell Square

*First published in 1968
by Faber and Faber Limited
London: 24 Russell Square
Printed in Great Britain by
W & J Mackay & Co Ltd, Chatham
All rights reserved*

S.B.N. 571 08502 4

Dedicated to Dr. Nathan Mutch, M.A., M.D., F.R.C.P.,
of Pitt White, Devon

Contents

Contents

Illustrations

LINE DRAWINGS

Preface

Over seventy years have passed since the last British book on hardy bamboos was published. It was written by A. B. Freeman-Mitford, who later became Lord Redesdale. His interest in the subject was aroused whilst he was serving as Secretary to the British Legation in Tokyo. His book *The Bamboo Garden* appeared in 1896, and although long out of print has remained a classic work of reference to this day. In 1914 the late W. J. Bean, whilst Curator of the Royal Botanic Gardens at Kew, published his authoritative work, *Trees and Shrubs Hardy in the British Isles*, and being especially interested in the hardy bamboo, included a charming account of its place in the Edwardian scene.

Many additional species have since been introduced into temperate countries, and a greater knowledge gained of their requirements and potentialities. To bridge this wide gap in the horticultural field the present book has been written. *Bamboos* is not designed as a botanical monograph on hardy bamboos, but primarily as a book founded on practical experience for those who are interested in growing these fascinating plants.

As Head Gardener at the pleasant streamside gardens at Pitt White in the village of Uplyme, East Devon, where over fifty different species and cultivars, perhaps the most comprehensive collection of hardy bamboos in Europe, have been established, I have acquired a special first-hand knowledge of the cultural requirements, characteristics of form and growth, and the various methods of propagation of these plants. Such observations, supplemented by information obtained from the inspection of other gardens, and by correspondence with other growers in other countries and in widely distributed zones throughout the British Isles, have enabled me, I hope, to clothe the bare bones of botanical record with life and practicality. For purposes of identification the familiar form of botanical key has been discarded in favour of a straightforward consistent description of the obvious features of each species, from the fresh new shoot to the mature cane and leaf, with a summary in each case of the details which serve to distinguish the species from others bearing a superficial resemblance to it—species with which it might easily to confused. It is hoped that this less formal approach will lead to more confident and definitive identification than has been possible in the past. It should be especially useful when applied to the members of the genus *Phyllostachys*, where many species are allied botanically, and may at first glance look very much alike. A table of synonyms and a list of the Chinese and Japanese names, plus a summary of the literary sources, are given at the end of the book.

It is, however, impossible for me to name all those who have helped me in correspondence or discussion, but I do wish to thank those individuals who by their efforts and special knowledge have enabled me to complete this work.

Firstly, I am deeply indebted to Dr. and Mrs. Nathan Mutch, the owners of the beautiful gardens at Pitt White, who have given me the freedom to spend a great deal of my time on the study of bamboos, and I wish to thank Dr. Mutch also for his valuable assistance in laying out the manuscript, and in checking the very complicated nomenclature. I must also thank Mr. W. D. Clayton of the Royal Botanic Gardens at Kew, who has patiently answered my many queries with unfailing accuracy and courtesy over the last three years, and many others. My good friend Douglas Haywood overcame great technical difficulties and secured many of the crucial photographs. The meticulous line drawings were executed with great skill by the gifted botanical artist Mr. R. E. Browne, of Uplyme, Devon. His idea to incorporate a blue tit in each of his drawings to show the comparative cane and leaf sizes was a radical but very welcome departure from the usual inch scale. The photographs for the frontispiece and for Plates 8 and 11, of the bamboos at Mount Usher in Co. Wicklow, Eire, and in the Botanic Gardens, Dublin, were supplied through the courtesy of the Irish Tourist Board.

I am indebted also to Dr. Hubbard of the Royal Botanic Gardens at Kew, Mr. D. McClintock, and Mr. Roy Lancaster of Messrs. Hillier & Sons of Winchester, for the many rewarding discussions we have had on the hardy bamboos, also to the Director of the Rosewarne Experimental Horticulture Station at Camborne, Cornwall. From the famous gardens at Bicton in Devon came valuable data, through the courtesy of Major N. D. E. James, O.B.E., M.C., Agent to the Clifton Devon Estates. I am grateful also to Commander Dorian Smith, owner of the world-renowned gardens at Tresco in the Scilly Isles, for much information concerning bamboos growing on the islands. Thanks to Mr. Cribb, Head Gardener at the Abbotsbury Gardens in Dorset, I was able to find the elusive *S. tessellata* and bring it back into general cultivation once more. The stands of various species of bamboo in the gardens are a tribute to his care and expert knowledge, and are worth a visit at any time of the year.

I have had much support from the Head Gardeners of many National Trust properties and private estates in many parts of the United Kingdom, especially in plotting the national distribution of the various hardy species.

From outside the United Kingdom much information came, and I am grateful to Mr. M. H. Gaskin, Officer in Charge of the United States Department of Agriculture Federal Experiment Station at Mayagüez in Puerto Rico, and to Mr. A. M. Kormilitzin, Head of the Department of Dendrology and Decorative Horticulture at the State Nikita Botanical Gardens at Yalta in the Crimea, U.S.S.R.

Eire, too, provided a wealth of data, especially from Mr. Aidan Brady, Assistant Director at the National Botanic Gardens, Glasnevin, Dublin, and Dr. J. G. D. Lamb, Head of the Agriculture Institute, Malahide, Dublin. Amongst the many owners of private estates who assisted me with practical information are: Viscountess Mersey, who owns the beautiful gardens at Dereen in Killarney, Co. Kerry; the Earl of Rosse, Birr Castle, Birr, Co. Offaly; Mr. R. B. Walpole, who has a varied selection of hardy bamboos on his estate at Mount Usher, Ashford, Co. Wicklow; Mrs E. Grove Annesley, the owner of a large collection in her gardens at Anne's Grove, Castletownroche, Co. Cork; Mr. R. Walker, whose bamboos at Rossdohan Island, Sneem, Co. Kerry, are probably the oldest established groups in Eire; the Rev. L. A. Robertson, Huntingdon Castle, Clonegal, Ferns; Mr. McKenzie, of Garnish Island, Glengarriff, Co. Cork; Mr. H. Dool, who has a good selection in his Winkfield Manor Nurseries at Waterford; Mr. Forrest, of the Gardens, Lisselane, Clonakilty, Co. Cork; and Mr. Killihan at the Bourne Vincent Memorial Gardens at Muckross, Killarney, Co. Kerry.

To all these and the many others who assisted me, may I give my sincere thanks, and trust that they will find increasing pleasure in growing hardy bamboos, and I hope also that the information contained in this book will prove to be helpful to them.

Introduction to the Bamboo

The bamboo is associated in the mind of the average person with the steaming hot jungles of Burma or Siam, or else with food for pandas. Throughout the centuries, due mainly to the mistaken supposition that all bamboos were tender tropical plants, a great deal of myth and legend accumulated round them. It was not until the middle part of the last century, when plant collectors of the Victorian era began sending back specimens to European gardens, that it was discovered that many of the species were, in fact, absolutely hardy. It had not been realized previously that the giant tree grasses were not confined to the jungles of the East. Some species come from North America, some from the wild and storm-swept tip of southern Chile, while others originate many thousands of feet up in the great Himalayan mountain range. Dozens of species come from areas where the climate is anything but hot and steamy. None, however, are native to Europe, although several species, particularly some of the hardy 'running' types of Arundinaria, have escaped from gardens and are now naturalized.

The origin of the word 'bamboo' is lost in the mists of the distant past. There is no mention of the bamboo in ancient writings until 400 B.C., when Ctesias, court physician to King Artaxerxes Mnemon of Persia, referred to it. No one can say with accuracy when the word began to be included in the spoken tongue. The answer to that question is still a riddle which puzzles etymologist and layman alike. Marsden records in his dictionary that it is Malay, but Garcia, in his *Colloquois dos Simples e Drogas e Cousas Medicinaes da India*, published in 1563, says when writing on the drug Tabasheer, 'the people from where it is got call it sacar mambum, because the canes of that plant are called by the Indians "mambu" '. Fitch came nearest to mentioning the word as we know it today in his narrative included in Hakluyt's *Navigations*, published in 1599 (Hakluyt II, 291), when he wrote that 'all the houses are made of canes which are called bambos, and bee covered with straw'. The English translation of Linschoten's *Navigatio ac Itinerarium*, printed in London by John Wolfe in 1598, mentions, 'a thick reede as big as a man's legge, which is called Bambus'. The word has now become integrated into the everyday language of most countries, and is generally accepted by botanists in every part of the globe.

The first hardy bamboo to be introduced into Europe was *Phyllostachys nigra*. Some authorities believe that the date was 1827, and others assert that it was much earlier than that. In any case, *P. nigra* (the black bamboo), arrived long before any of the other species, and was the vanguard of the dozens of hardy species which grace our gardens today. As more species are discovered every few years, I should not be the least surprised if the number of separate species under general cultivation in temperate countries reaches the hundred mark before the end of the present century.

This book has not been written with the tropical bamboo in mind, but I believe that it is worth mentioning some of the diverse uses to which this accommodating plant has been put in other parts of the world. In Asia the native uses the stout canes as structural members for his house building, whilst the fronds thatch the roof. The canes of larger diameter are cut and fashioned into food receptacles and cooking utensils. The succulent shoots are eaten with knives or chopsticks made from the twigs and branches. The dense wood, which is remarkably hard-wearing, supplies the native with his weapons and farm implements, with his pipe, his musical instruments, and toys for his children's play. W. Munro in his monograph of 1866 quotes authority for the observation that the natives of Malacca used to pierce the stems of suitable growing bamboos in such a way that flute-like musical notes were produced when the wind blew through them. Some such stems had a slit in every internode, and in a breeze would produce any number up to twenty different notes. The natives named these living orchestras 'bulu perindu', or plaintive bamboos. The bamboo is amazingly adaptable for a multitude of purposes, and wherever you look, in China, Japan, and other Asian countries, you will see the bamboo in use for one purpose or another.

Aside from the domestic and commercial uses of today, the bamboo has played its unique part in history throughout the centuries. The most famous instance of the bamboo helping man hardly needs repetition here; it was, of course, the smuggling of the silkworm eggs from China to the court of the Emperor Justinian of Byzantium in A.D. 552. Little did the two Nestorian monks know when they smuggled the eggs inside the bamboo staff that they would help to lay the foundations of the great silk industry in the Western world. The famous traveller Marco Polo recorded in the journal of his journeys through Cathay that 'the Chinese, since at least 200 B.C. had been using oil taken from wells drilled to depths of 3,500 feet. To enable them to do this, and reach these depths, the Chinese employed bronze drills enclosed in bamboo casings'. The use of the hollow bamboo tubes to carry the oil, is I believe, the first recorded application of the giant grasses to industry. He also recorded that the Chinese warlords used the bamboo for military purposes, and old prints can still be seen showing warriors using hollow bamboo tubes as a kind of primitive mortar to fire giant spears, and as rockets with gunpowder as the propellant.

In the field of medicine the bamboo had its uses. An ancient elixir, Tabasheer or Tabaschir, obtained from the branch cavities of certain tropical species of bamboo, was believed by Chinese and Buddhist priests to be the infallible cure for all ills. In India and China it is still used as a medicine, but has long been regarded by Western pharmacists as having little medicinal value. J. L. Macie (1791) described it as a siliceous concretion found in the nodes. W. Roxburgh (1819) found it in several species and almost filling the internodal spaces of *Melocanna bambusoides*. F. W. Mellor (1940) regarded it as a non-crystalline hydrated silica akin to the semi-precious stone opal. F. A. McClure (1966) states that it consists chiefly of amorphous silica in a microscopically fine state of subdivision, used nowadays as a catalyst, or catalyst carrier, to facilitate certain chemical reactions, some of commercial importance. Its ancient repute was as a curative for many conditions, and it is of interest to note that artificially produced amorphous hydrated silica has been used in recent years as an antidote for various internal infections of the food-poisoning type. Its mode of action is to remove toxic products by absorbing them into its own insoluble structure, where they are fixed and rendered incapable of causing further harm. It would seem that, after all, the traditional use of Tabaschir as a medicine may have had some logical foundation. Incidentally Tabaschir is not to be confused with the bloom found on the outer surface of the canes, especially in the zones just below the nodes. This varies in amount in different species, from a barely perceptible bloom to a fairly heavy covering of fine powder. This powder, which is particularly plentiful on a Chinese species, *Lignania chungii*, was analysed by Chang, 1938, and from it he isolated a number of not too complex chemical compounds; interesting amongst them was a substance closely related to one of the sex hormones.

The bamboo itself has been the subject of wood-carvers, and Plate 17 shows a bamboo pot carved on a three-dimensional plan by an unknown artist. The walls of the pot are barely an inch in thickness, yet the craftsman has succeeded in covering them entirely at all depths with a complex of patterns and figures in relief.

The fashioning of fine *objets d'art* from jade by the Chinese is universally and deservedly acclaimed. The means whereby the exquisite patterns and shapes are formed is, however, not so widely known. Jade, which is an extremely hard stone, is not cut by metal tools. Steel, even hardened steel, makes little impression on its surface. The soft lines and figures are gradually formed by the slow and steady application of an abrasive substance to the face of the stone. This wears away the jade to the desired shape. The abrasive is applied not by a metal tool but by bamboo sticks. These are readily available, light to handle, and although they are hard, are easily shaped to the angle required by the craftsman.

In Oriental literature and art the bamboo is depicted as one of the 'four noble plants', the others being the orchid, plum blossom, and the chrysanthemum. The bamboo is symbolic of the wise man who, shaken violently by the storm, bends but never breaks. In the eleventh century the bamboo was painted with great delicacy by Wen T'ung, and was the favourite theme for the paintings of the sage and poet Su-Tung-pô. Its popularity with the Oriental artist has continued throughout the years to the present day.

The article of dress shown on plate 18 is at least a century and a half old. Described as 'a nobleman's vest for summer wear in the Court of Japan', it is made entirely of tiny portions taken from the 'Hill country, male bamboo' (*D. stricta*).

The Bamboo Garden

---

*How to recognize the main groups—Bamboo hedges and
screens—Bamboos for the country house, the semi-urban garden,
and the patio or paved garden—Tub and pot bamboos*

It is remarkable how few people realize the potentialities of bamboos
as garden plants. They are cold resistant, evergreen, grow with varying
luxuriance as hedges and screens to protect more tender plants from
cold winds and frosts, and many species, especially the more decorative
types, make excellent specimen plants in lawns or neighbouring beds.
Most of the Phyllostachys, as well as members of the Arundinaria that
possess a close circumscribed rooting system, are suitable for decorative
plantings.

In selecting desirable species for any garden it is essential to bear in
mind the distinguishing features of the three main genera: Arundinaria,
Phyllostachys, and Sasa.

The genus Arundinaria comprises the largest number of species and
attempts have been made to subdivide it into several genera, but no
classification so far proposed meets with general acceptance even among
botanists, and for the present horticultural intent they can be ignored.

The word Arundinaria derives from the Latin 'arundo', a reed, and
gives the immediate clue to the characteristics of the genus. The stems
are round, smooth, and straight like reeds. They have no visible internodal grooves, and are partly encircled with branches. The branches do
not appear until the cane has achieved full stature, or nearly so, and the
lower branches are always the last to appear. Occasionally the central
branches open before the topmost ones, but the lowest are invariably
last. The cane sheaths clasp the cane tightly, and in some cases may
adhere to the branch or cane until desiccated and tattered with age. A
good example of this feature are the sheaths of *A. japonica*, which, when
forced from the cane by the opening branches, wrap themselves around
the base of the branch so tightly that they are only removed with some
difficulty. As a rule the rootstock of the Arundinaria is much more

active than that of the Phyllostachys, but with a few notable exceptions is seldom rampant or invasive. Most of the Arundinaria species are prolific cane bearers, much more so than the Phyllostachys, and this habit enables them to be utilized as hedging and screening plants.

The genus Phyllostachys gets its name from the Greek words 'phyllon', a leaf, and 'stachys', a spike, indicating a leafy spike. The lowest branches are the first to form, and the others open out serially from base to tip as in an inflorescence or spike of flowers. The canes are generally stouter than those of the Arundinaria, and show a marked tendency to zigzag from node to node. This feature is always present, although more striking in some than in others. The stem between the joints, or internode, is always flattened or grooved, often quite deeply on alternate sides. These indentations house the young branches, which when under the protection of their clasping sheaths, are pressed tightly into the soft pliable surface of the young growing cane. There are often secondary grooves in the hollows and these vary in number according to the number of branches borne, usually two or three at the corresponding node. The cane sheaths are not persistent, and are discarded very early in the growing season, although a few species, for example *P. aurea*, may retain their basal sheaths. The annual cane production of the Phyllostachys is much less than that of the Arundinaria, but the average basal diameter is greater. The rootstock is seldom free running, and the majority of species tend to grow in close tight clumps.

The third main genus, the Sasa or Saza (the word is believed to be a Japanese corruption of two Chinese words, Hsai-chu, or small bamboo), contains many hardy species which were formerly grouped under the Arundinaria. These are easily recognized by their dwarf habit of growth, and by the size of their leaves, which are usually broad in comparison with the height of the slender canes. The branches are invariably long, and borne singly or in pairs. The cane sheaths are very persistent, and the rootstock active, often extremely so. The Sasa when once established form dense thickets, and thrive even where there is overhead cover, this being their native environment as vegetation of the forest floor.

With their wide range of cane colouring, their delicacy and variations of foliage, most species of hardy bamboo can be planted as specimen plants of great ornamental appeal. The selection of a species for a given site requires careful consideration. Obviously it is unwise to plant a rampant-growing species in a restricted area, and foresight is needed to visualize just how the original plant should look in six or seven years' time. The background is important, too, and imagination must be used to see the mature plant in every season of the year. In the dead of winter, when all deciduous plants have shed their leaves, it is really surprising how tropical an effect can be created by a few clumps of bamboo.

All bamboos thrive in moist and mild environments, but the hardy species can be grown successfully in temperate climates, and in a wide variety of conditions, provided that care is taken when they are first installed. They will not thrive in a windy exposed position unless protection is provided. Winds from the north and east are particularly harmful. A south or south-west slope is ideal, providing the maximum of warmth for the longest period of the year. Gardens possessing such a setting are comparatively rare, but fortunately hardy bamboos can be induced to grow in many less favourable circumstances if shelter is provided during the first season, and possibly during the second if the species is a reluctant Phyllostachys. In later years the clump will thicken and the accumulating older canes will form a tough dense thicket to protect the mat of rhizomes which lies just under the soil surface, and will give shelter to the soft new spears emerging during the early spring. Shade is important, too. Certain species revel in full sunlight, others must have shade, but on the whole most grow better where there is at least semi-shade of some kind. They prefer the kind of dappled light found in woodland fringes.

The hardy bamboos range in size from pygmies of less than a foot tall to giants of 30 feet or more. The canes present a wide range of colour, especially amongst the Phyllostachys, and no two species are exactly alike. The colour varies through many shades of green, to brown, purple, black, and gold. Some are spotted or blotched with contrasting colours, whilst others have a combination of different colours in their make-up. There are those which boast silver, yellow, orange, or golden variegations in their foliage. Added to these singular qualities is the fact that they are evergreens, and keep their bright, rustling foliage throughout the year. It is true that the hardy bamboos introduced into temperate climates from warmer countries seldom reach the full height and majesty attained in their native homeland, but this is only to be expected, as the weather, particularly in the United Kingdom and in north-west Europe, is too erratic, and long periods of warmth combined with high air humidity do not often occur. Nevertheless some species can make canes of fairly imposing heights in the milder districts. An owner would indeed be hard to please who could not find some suitable species of hardy bamboo to grow, no matter whether the garden has a naturalistic setting or is more formally planned.

In larger gardens every horticultural possibility of the tribe can be enjoyed—as windbreaks, visual screens, specimen plants, and ground cover. As a visual screen, *A. anceps* is more attractive than *A. japonica*, and is usually appropriate unless the ground is so moist and fertile as to promote an excessive spread of growth. For shady places *A. nitida* is more suitable. It is non-invasive and has a particular delicate charm, but must not be too much exposed to the wind. As a windbreak, *A. japonica* is the choice. In 1956 a hedge of this species, a hundred yards long, was planted at Rosewarne Experimental Horticultural Station in Cornwall, and is now an effective barrier against the Cornish gales. Four years after planting many individual canes stood 8 or 9 feet high. In another trial four hundred species of hedging plants were carefully compared and assessed for possible agricultural use as shelter hedges, and *A. japonica* was one of the eight finally selected as most suitable. Such hedges can grow to a useful height and any unwanted spread can be controlled by rotavating, or by the removal of the soft shoots as and when they appear. Wind velocities are reduced by a solid barrier for as far as thirty times the height of the obstruction, but the degree of protection needed for ordinary garden purposes probably does not extend farther than quarter to one-third of this distance. It has been found that a hedge presents a more satisfactory barrier than close fencing, stone, or brickwork, etc. A solid wall, ironically, gives rise to air turbulence on the leeward side, and the swish and eddies of the disturbed air do more damage than would have been done had the wall never been erected. Experimentally it is shown that the most satisfactory barrier is one with a 50–60 per cent degree of permeability. A bamboo hedge meets these requirements.

A purely bamboo garden could easily be devised, but it is more usual to establish these plants in attractive combination with shrubs and trees. To illustrate various selections and uses, three different types of garden can be considered in more detail. Firstly, the country house garden with differing slopes and levels, the shade of mature trees, and possibly a little water in the form of a pond or small stream. Then the semi-urban garden, more restricted in size, and with less variety in its flatter terrain. Lastly, the paved garden found as an immediate surround to a country house, or as a small free space behind a house in town, or as a forecourt or terrace of a larger hotel, municipal centre, or block of offices or flats.

The country-house garden may require a hedge, and so may the smaller type of semi-urban garden, so we will begin by planting the hedge. This can be grown either from divisions taken from an established clump or from rhizome cuttings. If the former are used, they should be dug in roughly 2 feet apart, but if there are ample supplies of planting material plant each piece as close to the next as is possible. Close planting enables a hedge to become effective quickly, as the rhizomes and roots intertwine and the ensuing canes press tightly together to make an impenetrable barrier in a couple of seasons. A less costly hedge can be grown from single rhizome cuttings with one or two canes attached. These, too, should be planted 2 feet apart, but each cane must be staked firmly to stop the rhizomes being loosened by the wind. An even cheaper method of hedge growing is by the planting of simple rhizome cuttings. The last two methods, and especially the latter, take some years to grow canes in sufficient quantity to fulfil the purpose, but they will, in time, make just as good a hedge as the one obtained from large divisions. It is possible to forecast roughly the future density of the newly planted hedge by doubling the number of canes produced each season. A first-year single rhizome may produce two canes in the first year, four in the second, and so on.

A trench some 6 inches deep and one spade wide is taken out of the prospective hedge line. Coarse sand is sprinkled on the bottom of the trench, deep enough to cover the cuttings. The rhizomes to be used must be taken from a healthy parent plant. They must also be short jointed and furnished with plenty of growing or embryo cane or root buds. Cut the rhizome into 6-inch lengths, and place them 6 inches apart along the length of the trench. Place them in alternating positions, that is, one placed vertically and the next horizontally. The reason for these alternate planting positions is that cuttings placed in a horizontal position tend to make roots rather than shoots at first, but when they are planted vertically the reverse happens. So by planting the cuttings in alternate positions along the trench one is able to have both roots and shoots quickly along the length of the hedge line. Hedges have been grown by layering whole rooted growing canes in a trench, but only a few species are suitable for this purpose. As a rule the results are not encouraging and this method cannot be recommended.

Regarding plant protection and shelter, I should mention here that hurdles made from bamboo give good results. They are portable and simple to make. Whole or split canes woven into a rigid framework can be manufactured in a matter of minutes, and can be made to fit any size of plant. The thin pliable canes of *A. nitida* are ideal for this purpose.

To return to the bamboo hedge. Every gardener knows that there is no such thing as an instant hedge, and a bamboo hedge takes roughly the same time to acquire a useful density as do privet, laurel or quickthorn. Nevertheless it is superior to those in many important ways. It quickly attains height, and does not require an annual pruning to promote its final density. The mature canes only need to be cut back if it is wished to limit the height of the barrier, but such cutting will not increase the density in the lower parts which are retained. Its rooting system does not extend laterally to any great extent beyond the area of visible surface growth, and therefore, unlike the common hedging plants such as privet, it does not impoverish near-by beds by robbing the soil of its nutrients.

An interesting booklet written by T. U. Hartwright, called *Planting trees and shrubs in gravel workings*, has been issued by the Sand and Gravel Association of Great Britain. In it are listed the many different plants which can be used to provide screening or add artistic improvements to existing gravel pits. Amongst the plants recommended are six species of hardy bamboo. These are *A. fastuosa, A. japonica, A. nitida, P. henonis, P. mitis,* and *P. viridi glaucescens.* The author emphasizes the suitability of the hardy bamboo for screening purposes, and states that their adaptability is scarcely appreciated at the present, and I entirely agree with him. What is more, useful garden canes can be cut from the hedge . . . no other hedge has all these advantages.

Formerly *A. simonii* could have been used instead of *A. japonica,* but this species began to flower last year, and it is unlikely that sufficient material for planting will be available for many years to come.

Now that the hedge has been planted we will look at the other positions where bamboos can be planted. Under shade trees there may be, and usually is, a blank void. Very few plants like dense shade; even grass despises it. Some species of bamboo, however, grow best in such a site, and indeed some species prefer overhead cover. The graceful *A. nitida,* with its delicate, trembling leaves suspended on thin, arching, whiplike canes, loves shade. The paper-thin leaves are intolerant to the direct blaze of the summer sun, and will curl up at its first touch. They do, however, open out again to their normal shape as soon as the sun goes behind a cloud. Several species of bamboo prefer the shade, but when the situation calls for an elegant species, *A. nitida* must take precedence of all others. If the area is rather wild, or if the bamboo is needed only to fill a space and have no particular appeal, then any of the Sasa species can be planted as ground cover.

Every large garden has a neglected corner, or an out-of-the-way section where nettles and docks run riot. How many times have we heard it said, 'I simply must clear up that corner one day', but that day seldom arrives. In such a corner a few spaces can be cleared and two or three clumps of rampant-growing species planted, such as *A. vagans, A. humilis,* or *S. palmata.* These species do not take long to overrun the area, spreading as they do by their very active rhizomes. The umbrella of dense foliage deprives the weeds of light, and in a short time instead of the tangled mess of weeds there will be a close thicket of bamboos. It must be stressed, however, that these species *must* only be planted in out-of-the-way plots. If a less rampant-growing species is desired, *A. marmorea* or *Bambusa quadrangularis* can be substituted.

We come now to the principal part of the garden where ornamental bamboos can be planted, namely on or near to a lawn, and by the side of a stream or pond. A great range of species is available to choose from. A lawn ornamental bamboo must be graceful, clump forming, but not so arching in habit as to interfere with lawn maintenance. The Phyllostachys are usually to be preferred to the Arundinaria, for not only are they generally clump forming in habit but they also display a greater variety of cane colouring. The coloured cane species are seen to their advantage when planted in a position with a grassy slope or a dark background of shrubs or shade trees. Planted as specimen plants, bamboos do not rob the grass of food; in fact, grass will grow right up to the edge of the clump, but it is better to keep a ring of open free soil around the perimeter of the site so that the grass can be cut neatly at the edges. It is also necessary to feed ornamental clumps annually with a rich top dressing. The group must also be checked over periodically to remove dead or damaged canes which would spoil the symmetry of its shape. Outstanding amongst the Phyllostachys species for ornamental site planting are *P. aurea,* the 'bamboo of fairyland', which has golden canes when grown in a sunny position and has distinctive swollen joints on the lower part of the canes. *P. castillonis* and *P. sulphurea,* two other golden-caned species, are also well worth planting, although the former may be hard to come by since flowering profusely in recent years. Also recommended are *P. mitis,* which is somewhat similar in habit to *P. sulphurea,* but has green canes, *Chusquea culeou* from South America, a solid-caned, impressive species which bears short whorls of branches covered with tiny leaves, and the tall, slender, dark green *P. viridi glaucescens*.

Although I have stated that few of the Arundinaria species make good specimen plants, there are two that make good lawn groups, but can only be grown in mild areas. They are *A. falconeri*, the 'noble' bamboo, and *A. hookeriana*, the bamboo which has most conspicuous attractive variegations in its cane colours.

Just as the erect-type bamboo fits into the lawn scene, so a site by the side of a stream or pond is complemented by the weeping or cascading forms. The bamboo is in its element when planted near water, but it abhors boggy conditions. It must be planted so that the rhizomes are well above the water-level. The *puberula* group of the Phyllostachys, with their graceful arching canes and tumbling masses of medium-length foliage, impart a true Oriental flavour to the waterside scene. Of special merit in this group are *Phyllostachys nigra*, with its ebony-black canes, the brown-and-gold-caned *P. fulva*, the golden-caned *P. henonis*, and the black-speckled canes of *P. punctata*. *P. boryana*, with its black-splashed golden canes, is considered by many to be the most attractive of all the cascading types. Plate 12 shows a group of the common *A. japonica* growing by the side of the stream at Pitt White, and it does look attractive, but tends to look untidy unless the canes are thinned out regularly. Whilst on the subject of streamside bamboos I would like to mention that some of the semi-dwarf rampant-growing species can be used to stabilize loose banks and screes. Their tight network of slender but tough roots and rhizomes bind the loose soil and prevent erosion. *A. vagans* and *A. humilis* are recommended for this work.

The smaller semi-urban garden, restricted by lack of space and usually by the flatter site and drier soil, has a much smaller range of species from which selections can be made, but the tall-caned types can still be used in positions usually allotted to flowering cherry trees or crabs. For boundaries, hedges of *A. japonica* are suitable within definitive fences or walls, with *A. nitida* where shade is cast by overhanging trees. *A. anceps* will also be suitable if the soil is sufficiently dry to limit its overgrowth. Along roadside boundaries, cascading types should be avoided, since they could hang too far over the pavement when wet with rain. Of course, there is always the risk that small boys may covet the canes for fishing rods, so that some kind of fence or wall should separate the bamboos from the road, and only slender-caned types such as *A. nitida* should be planted near the entrance, where they are needed to secure privacy and deaden the noise of heavy traffic. For deeper visual screening *P. viridi glaucescens* can be considered. The pattern of planting in the garden proper can be enriched by arranging in suitable apposition the erect and the contrasting cascading types. The erect lines can be secured sufficiently well with *A. spathiflorus*, *P. aurea*, *A. nitida*, and *A. murielae*, and the cascading or widely arching element furnished by the members of the *puberula* group already mentioned or by *P. flexuosa*, or, for example, a group of *A. nitida*, *A. fastuosa* and *P. aurea* could provide a graceful focal point when planted in the lawn or nearby a semi-rough patch or border. When a selection has been made the plantings should be kept at least 12 feet apart to avoid premature merging.

Bamboos should rarely be introduced near the walls of the house. It is not that the rhizomes will damage the wall, but they can find their way into surface drains and block them as they expand. Moreover, bamboos do not grow well in dry places, and the soil close to a wall is often arid.

Of the dwarfer varieties, few can be planted without the risk of their spreading beyond the areas allotted to them, but *A. marmorea* can be advised as pleasant ground cover for a deserted corner. *A. auricoma*, with its attractive golden-striped foliage, and *Shibataea kumasaca* can be safely introduced into the beds.

The treatment of a terraced area or patio near the house can be dealt with in the same way as paved gardens in general. The beds of soil in such areas are usually relatively small and strictly circumscribed, taking the form of small plots or borders let into the pavement surrounds of stone, concrete, or tile. Additionally there may be raised beds of more architectural form, and large tubs and containers of wood, cement, fibreglass, etc. Grown under these conditions the height of growth, which depends on the size of the underground rhizomes, will be less than in plants grown in the open garden. They will thus adapt themselves agreeably to the restriction of their more confined environment. It is advisable, nevertheless, to exclude species with long vigorously growing rootstocks and confine the choice to those likely to form compact clumps. As specimen plants for prominent positions, probably the easiest form to grow would be *A. nitida*, which is popularly used in this way in Copenhagen and other places abroad. But *A. murielae* and *A. spathiflorus* would be equally suitable, and even *P. aurea*. All these have an appropriate semi-upright mode of growth. If a species with moderately overhanging branches can be accepted, then *P. viridi glaucescens* would be appropriate. The *puberula* group (*P. nigra*, *P. henonis*, etc.)

would only be suitable where a considerable degree of overhang is needed. In urban conditions their growth would be of reasonable luxuriance. As a border plant, the silver-striped *A. fortunei* is attractive. It is a dwarf, standing perhaps a couple of feet high, and has an active rootstock; this being so, it would be unwise to use the bed for any other purpose. The golden-striped *A. auricoma* could, however, be safely used in small plots in association with other plants.

For tubs and large bowls, tall species with running rootstocks are unsuitable, because their rhizomes soon extend to the margins and coil around the inside of the container like the roots of a pot-bound plant. The most appropriate bamboos are the types which form clumps by a close gradual outward spread; for example, *A. nitida*, *A. murielae*, and *A. spathiflora*. In mild districts where there is space, and more care can be given, *A. falconeri* and *A. hookeriana* can be used effectively. The containers can be 3 feet or more across, and as much as 2 feet deep, but if some degree of portability is to be maintained they should not greatly exceed 18 × 12 inches and 18 inches deep. Containers of this size can be made of Green Cuprinol-treated wood, and fitted with two strong handles. In small boxes of this kind almost any bamboo which does not form stout long rhizomes can be grown attractively for many years and will produce canes of limited height and correspondingly narrow diameter. Ample provision must be made for good drainage by the use of crocks and the provision of holes in the base of the container. Regular watering, too, is essential, but perhaps not more regularly than would have been needed had the boxes contained herbaceous plants or half-hardy annuals. During hot weather an occasional syringing with water is beneficial. This helps to restore leaves which may be flagging, and to remove deposits of dust and smoke.

Regarding the soil for the boxes, any ordinary garden soil will suffice, provided that it is enriched with some type of compost. An annual top dressing is also desirable to keep the plants healthy and happy.

The many species of hardy bamboo grown at present in Europe and North America were for the most part introduced during the latter part of the last century. Since World War II little new material has been brought over from abroad, although many species wait to be tested for hardiness. Some of the lesser-known species, such as *A. amabilis*, *A. indica*, *P. nidularia*, are already growing out of doors at Pitt White and seem to be quite hardy. Most of the popular species were imported by private garden owners during the heyday of the bamboo in the

Victorian era, but since the break-up of the larger estates after the two great global wars the activities of the amateur plant collector have been confined to the acquisition of smaller plants, such as alpines. It is now up to the municipal and larger commercial growers to introduce new hardy species with decorative appeal. It is true that the phyto-sanitary regulations are severe, but this is necessarily so, and applies to all plants imported from abroad. In spite of this, the trouble and the risk involved should be well worth while in the interests of horticulture. Of course, a wide selection of hardy bamboos is already available from some leading nurseries, where care has been taken to preserve and propagate them, but the number of such collections is, alas, small. A list of nurseries which supply hardy bamboos is given at the end of the book, but has no claim to be comprehensive.

It is my earnest hope that by writing this book I can pull away the exotic blanket which has enveloped the bamboo, and release the plants into the open air of every garden. It is not so very long ago that many plants brought into Europe from foreign lands—plants such as the azalea, rhododendron, and even the tomato—were regarded with as much suspicion. A. B. Freeman-Mitford, later Lord Redesdale, long ago remarked on this tendency to think of 'foreign' plants as requiring special knowledge for their cultivation. His attempts to introduce the bamboo into the open garden brought sneers of derision from professional gardeners, who decided that his plants could never survive the cold winters of Britain. Despite their gloomy forecasts, the hardy bamboos are now well established, and some species are, to all intents and purposes, naturalized. They grow freely all over the country, from the north of Scotland to Ireland, and in every county in England, and each clump stands as a memorial to that great man's foresight, insistence, and courage.

✦ 2 ✦

How the Bamboo Grows

The growth cycle from rhizome to leaf and flower

All species of hardy bamboo have certain points of similarity. They are all evergreens. They all possess hollow-jointed rhizomes or underground stems, which radiate from the base of the parent plant to greater or lesser distances according to the species and the local conditions of soil and climate. They also bear aerial stems, or canes, which are hollow-jointed, too, and carry the branches and leaves. The bamboo of the tropics grows faster than any other living plant, thrusting its hollow-jointed stems higher and higher. In India, a tropical species, *Bambusa tulda*, is on record as having grown $1\frac{1}{2}$ inches in one hour. In North Africa, *P. mitis*, a hardy species, is known to have achieved the phenomenal growth rate of 20 inches in the space of twenty-four hours. The hardy species which grow in our gardens naturally do not grow as fast, as their growth rate is only intensified by the warmth of the summer sun; nevertheless, in the main growing months (June, July, August) I myself have measured shoots of *A. niitakayamensis,* Pitt White clone, which have made 6 inches of growth in twenty-four hours. As the shoot makes the great part of its growth during the daylight hours, and is more sluggish at night, it does not need a computer to work out the fact that a shoot from one of the taller-growing species can grow into a 30-foot cane in three months or less. I have seen hundreds of such shoots grow at that pace, day by day, and I am still amazed at the fantastic speed of growth that the bamboo achieves. From the sharp spear-tipped point, just showing through the ground, to a towering cane in full leaf in three months certainly beats the speed record formerly held by the bracken growing on the Scottish island of Ulva (15 feet of frond in a single summer).

The cane, when at the height of its first season's growth, acquires basal roots, and presently the rhizomes give birth to buds which expand into fresh rhizomes to store the food substances coming from the leaves, and explore further afield to establish next season's new colonies of canes. Thus the bamboo spreads.

29

A typical rhizome is an ivory-coloured underground stem, which has joints separated at intervals by hollow internodal sections. The septum, or joint, which seals off each section of the rhizome into compartments, is the focal part of the bamboo's growth, for from it spring the rooting system and the buds for the production of stem and new rhizome branches. The nutritional solutions carried along the internodes from the leaves and the roots of the parent plant are concentrated in these joints, and stored there for further use. Each joint is wrapped tightly in its own protective sheath. These sheaths are seldom seen, since their life is short, being terminated quickly as the ring of roots thrust themselves through the thin membranous covering. The stem buds lie snugly in a horizontal position at the base of the joint. These often remain dormant for a long time, and may not even swell until stimulated into growth some twelve months or so later in the following season, when they turn upwards towards the surface.

The tip of the rhizome is sharp and hard as horn to enable it to force its way through the solid soil on its journey to the outside world. If during its journey underground the tip is thwarted in its forward path by a solid object such as a large stone, the tip will often change direction and emerge prematurely from the ground, loop over the obstacle, then resume its journey below ground. This it continues to do until the limit of the season's growth has been attained; then the tip turns upwards, breaks through the soil to begin another form of life as a terminal aerial cane.

If the surrounding soil is dry, one can often discern where the shoots are going to appear, by the circle of moist soil which precedes the arrival of the shoot. This damp circle is formed by beads of moisture seeping from the higher sheath tips of the shoot, softening the soil immediately above the tip, thus easing the passage of the shoot upwards. This interesting phenomenon can be observed at close quarters, and could incidentally provide a new aspect of plant behaviour for observation by school botany classes. All one needs is a deep glass tank, some soil, a quantity of sand, and a portion of bamboo rhizome. Plant the rhizome in the soil at the bottom of the tank, and at one corner place a piece of piping, long enough to reach the top. Fill the rest of the tank with dry sand, and water through the pipe. This allows enough moisture to wet the soil, but keeps the sand dry. After a period of time the shoots can be seen swelling and reaching upwards towards the surface. As the shoot reaches through the sand one can observe the dry sand gradually becoming darker as the moisture from the shoot

B. quadrangularis. Rhizome and Rooting System

reaches it. Of course, any kind of container will do for this experiment, but if a glass container is used the student can see just how the shoot develops from the bud on the rhizome.

The emerging pointed tip of the rhizome is wrapped in a thick coat of overlapping sheaths which protect the branch buds nestling closely at the joints. By the time the eighth or ninth sheath is exposed above ground-level, the diameter of the shoot at that stage is the diameter of the cane to be. Each successive cane appearing in the group will be of larger diameter than its predecessor, until the optimum thickness of the cane of that species, compatible with local conditions, has been reached. Cane height is related to cane diameter, and as each successive cane emerges slightly larger in diameter, its height is found to be greater, too, until the average height of the species has been achieved. The diameter does not increase as the cane ages, neither does the height increase after the end of the first season's growth. So, unlike trees or shrubs, the bamboo cane remains for its entire lifetime at the measurements it achieved during its first season of growth.

After three seasons the cane reaches maturity, and may live for up to ten years or more. In the course of time the wood changes gradually from its original soft pithy consistency to a dense hard wood enclosed under a thin but hard surface skin. In some cases the tough epidermis contains so much siliceous material in its make-up that a match will ignite when struck upon it.

As silica features so much in the make-up of the bamboo, I feel that I must explain its role in botany. In some plants it is the principal mineral constituent. For example, the ash of *Equisetum* (the common horsetail) may contain as much as 80 per cent of silica, and it may constitute up to 40 per cent of the ash of wheat straw. Its exact role in the biochemical economy has not yet been clearly defined, but it has long been established that silica cannot be regarded simply as a substitute for cellulose and lignin, the basic materials employed to confer essential rigidity on a plant. F. Sachs (1856) demonstrated that silica-free straw obtained by special cultural methods was just as strong and rigid as normal straw of the same species. Concentrations and deposits of silica are usually greatest in the surface layers of the stems and leaves, and in the bamboo accumulations may be found near the nodes. In many species small intra-cellular silica bodies of various shapes and sizes can be seen in the thin layers of the foliage, and cells which are completely filled by such deposits are called silica cells. The precise location and distribution of such particles and cells has been used by the

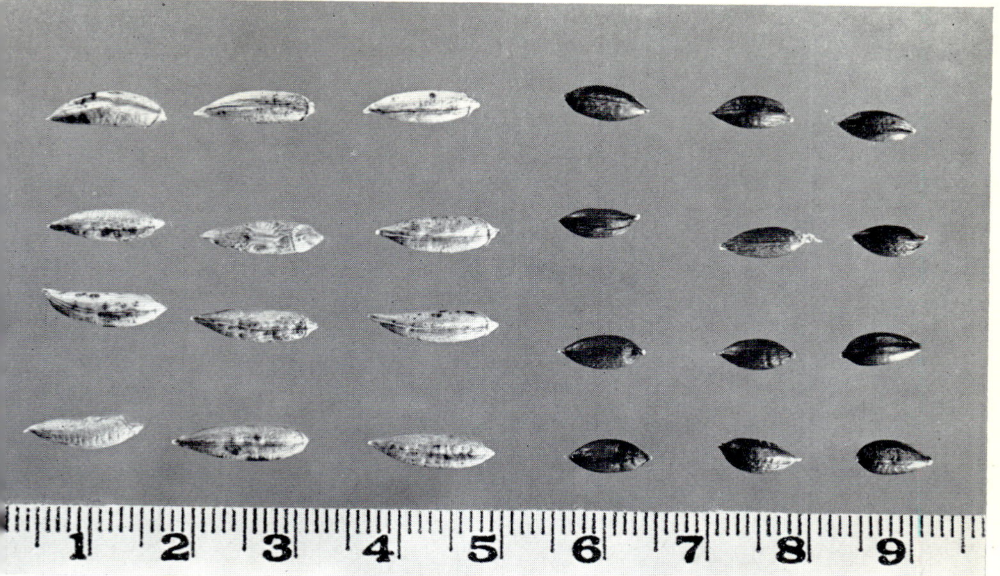

1. Seed from *A. simonii* (left), and *S. palmata* var. *nebulosa*. Collected from Pitt White, Sept. 1967

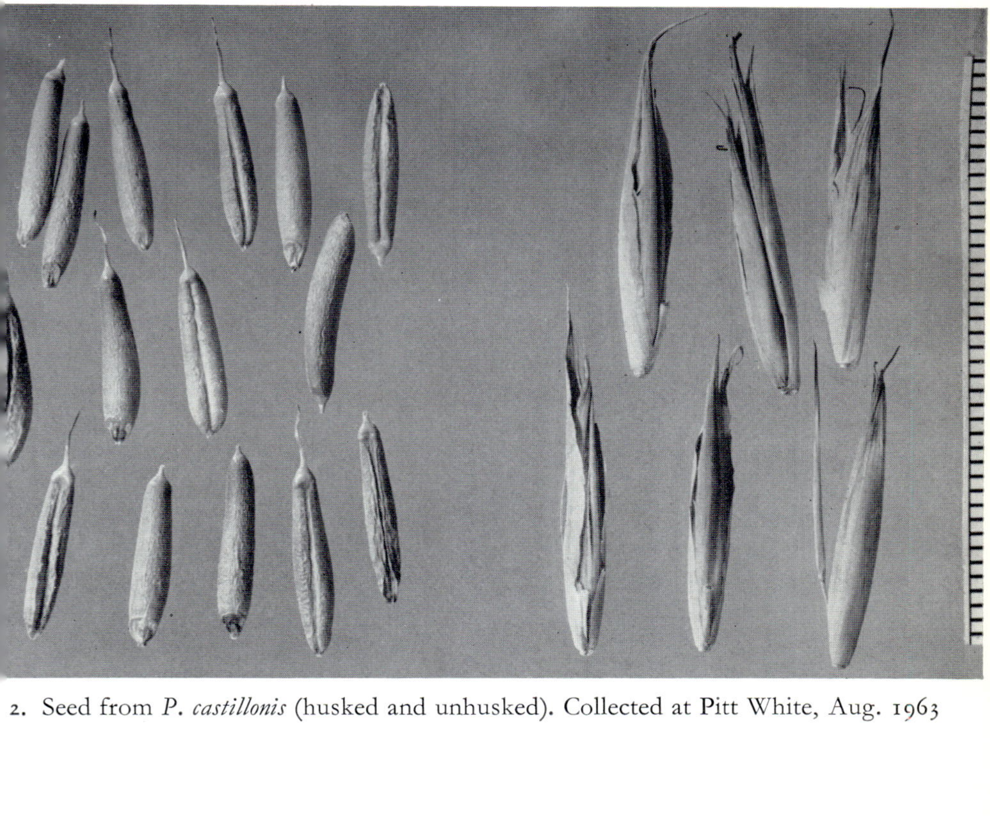

2. Seed from *P. castillonis* (husked and unhusked). Collected at Pitt White, Aug. 1963

3, 4.
One-year-old seedlings
grown from seed collected
at Pitt White from
P. castillonis, 1963

Japanese botanist K. Ohwi as the basis of a botanical key for the differentiation of various genera and species, but this does not seem to be sufficiently definitive to be of much service at present.

Many species can be identified by the characteristics, such as shape, tint or colour pattern, of the sheaths on the new shoots. Often species which are closely related and difficult to distinguish apart when fully grown, can be differentiated by the colour of the shoot sheaths alone.

Each sheath consists of three separate parts:

(1) *The sheath proper,* which may be marked by blotches or streaks of colour. It may also be smooth in texture, striated, or covered with adpressed hairs. The interior surface is often highly glazed, and marked with grey or black spots.

(2) *The limbus or sheath blade,* which is a short leaf-like appendage growing at an angle from the blunt tip of the sheath. Sheaths on the higher parts of the cane very often carry blades which develop into true leaves, whereas those on the lower portions may only be scales. The blade is narrow and strap-shaped, usually green in colour, and generally short-lived.

(3) *The ligule,* which is a slight curved rim-like projection along the inner margin on the blunt tip of the sheath. It may be fringed with hair, and flanked at each end by a small prominence (the auricle or ear). The auricles vary in shape and size, may be single or paired, and carry hairs or be smooth; characteristics which again are useful as points of identification of the species to which the bamboo belongs. The *ligule* is in effect a protective barrier between the blade and the sheath, its primary function being to prevent the entrance of moisture to the soft wood of the new cane.

As the cane is drawn upwards, the sheaths, fixed as they are below the base of the nodal ring, leave exposed an ever-increasing portion of the cane surface. Upon this surface can then be seen a waxy white bloom which is yet another of Mother Nature's waterproofing devices to protect the soft wood of the young cane. In some species this deposit is quite thick, deep enough, in fact, to be scraped off with the finger-nail. In others the coating is so thin as to be almost invisible. This substance is not the 'tabaschir' of the medical world, but is simply a straightforward waterproofing device.

When the branches burst free they force the restraining sheaths away from the cane. Some may hang on persistently for a time, often wrapping themselves around the branches before they part, but mostly, their task completed, they fall discarded to the ground. The branches and circling nodes now lie exposed. Each node or joint marks by its projecting rims the outer limits of the inner sealing septum. Some nodes stand out prominently, others are so inconspicuous that they may hardly break the outer contours of the cane.

The branches vary in number according to the species, from one to a dozen or more. On the Arundinaria the lower portion of the cane is usually devoid of branches, whilst the larger number always come from the central portion. The branches are borne, as they are on all types of bamboo, on alternate sides of the cane. Each branch divides into numerous branchlets which carry the leaves. On the smaller-leaved species the branchlets subdivide again to become little twigs which carry the myriads of tiny leaves. These branchlets and twigs remain on the main branch year after year, continually dropping the ageing and yellowing leaves and leaf sheaths and renewing with fresh foliage. All hardy bamboos are evergreens, and as with most evergreens this renewal is not a sudden occurrence, but receives its greatest impetus with the arrival of the mild spring weather.

The leaves are borne at the ends of the twigs and branchlets in alternate series, the spacing varying with the length of the branch. Although they vary in size, the shape remains roughly the same in every species: linear or linear lanceolate. The tips are always pointed, some more acutely than others, and the base is usually rounded or gently bevelled off to an obtuse angle. The bamboo leaf has a sheath just like the other grasses. The stalk, or pseudo-petiole, between the leaf-sheath and leaf-blade, and its ability to break away from the top of the sheath, distinguish the bamboo from the rest of the grasses. Other members of the grass family bear stalkless leaves that attach themselves by wrapping their lower parts around the stem.

The leaf-stalk is the basal extension of the midrib which bisects the leaf as far as the tip. The midrib may be conspicuous and palpable, or it may be fine and threadlike. It is the main collecting channel for the food solutions manufactured by the photosynthetic action of the leaf.

Running parallel, at set distances, between the midrib and the leaf margins are the secondary veins, which are finer in structure than the midrib, and if one looks closely between the secondary veins with the aid of a strong handglass, one can often see another set of even finer veins known as the intermediary veins. Both sets of veins appear in pairs on each side of the midrib. The number of pairs per leaf is another useful means of identification of a species. The space between the veins is occupied by a series of fine longitudinal lines, and connecting these fine lines are the cross-channels which break up the pattern of the leaf into small oblong or roughly square-shaped sections somewhat resembling mosaicwork, or tessellation, as the feature is botanically known.

The leaves of *all* hardy bamboos are tessellated, though the feature may be more conspicuous in some species than others. It has been stated that the clearer and more conspicuous the tessellation, the hardier the species. This is speculative to say the least, since there are a number of tropical species which show leaf tessellation, but which cannot tolerate even the slightest drop in temperature. The generalization does, however, represent a partial truth in that all our robustly hardy types have well-tessellated leaves, and all our non-tessellated species require either greenhouse treatment or specially selective warm growing sites.

The leaves of the non-tessellated species differ in being dotted with numerous translucent glands, and in the greater prominence of their secondary and intermediary veins.

Some species of semi-hardy bamboos which are of the non-tessellated group come from the slopes of the high Himalayas. This may suggest a paradox, but may be readily understood when it is realized that in their country of origin these species do not grow in the familiar form of the bamboos as we know them, but take on the habits of herbaceous perennials, sending up separate leaf- and flower-bearing canes every year which die down each winter to reappear in the following spring. The fresh canes and greenery spring from the hardy rhizomes which have remained undisturbed and snug beneath the snow-covered ground. However, when these species are grown in warmer climates the canes do not die down during the winter, unless it is exceptionally severe, but persist long enough for the wood to harden into true canes. Each year the succeeding crops increase in size and girth until they bear little resemblance to their original parents. They also lose the habit of bearing separate leaf- and flower-bearing canes.

Look closely at the leaf edges of a bamboo leaf and it can be seen that the margins themselves are entire, i.e. they are not dentated or toothed, as has often been stated. Leaves may be seen which possess wavy corrugations, but this is usually an immature feature which disappears as the leaf develops. Attached to the leaf edge there is frequently a line of minute bristly hairs; sometimes they appear on one margin only, but usually on both. These bristles may be so small as to be almost invisible to the naked eye, but when present they all slope in the same direction—at an acute angle towards the tip of the leaf. The purpose of these bristles has yet to be fully explained. They may be a means of directing water away from the base of the leaf towards the tip, and so to the ground, or they may be the remains of rudimentary spines. To detect them when seemingly invisible, try drawing the edge of a leaf, from tip to base, along the upper lip. When the bristles are present they produce a sensation like that of a fine saw blade.

The upper surfaces of the leaves are invariably smooth and glistening, and after a shower of rain they flash and flicker like mirrors when the sun catches them. The under-surfaces of the leaves are always sombre in colour and often coated with minute hairs. The dull matt effect is produced partly by innumerable minute peg-like projections protruding from the lips of the many stomata, and are designed to prevent globules of moisture from adhering and choking the entrances of the tiny breathing passages. Immerse a leaf in water and it will be seen that whilst the upper surface retains its smooth and glossy appearance, the underside acquires a glistening silvery coating, formed by the numerous bubbles of air trapped and held by the pegs. The leaf can be left immersed in water for a considerable period, and even agitated from time to time, but the pegs will continue to hold on to the air bubbles, and will do so until the cell structure of the leaf begins to decompose. This simple experiment could provide an interesting discussion point for school plant biology classes.

The individual flowers of the bamboo are small and inconspicuous, and have only one ovary with one ovule, giving rise to one seed or grain per flower. They are arranged in threes or fours as spikelets, which are then grouped together as spikes, racemes or panicles, like the flowering heads of cereals or grasses. Although the individual flower may be as insignificant as those of any grass, they form objects of beauty and grace when arranged in their panicles and spikes. *P. castillonis*, with its fine flexible stalks, produces an enormous complex of panicles borne on canes 15 feet or more in length, resembling giant loose pendulous sprays of diminutive oats which tremble at the slightest breeze. *A. fastuosa*, by contrast, flowers in groups of compact spikes more reminiscent of wheat; see Plate 6.

The act of flowering is not an annual one, and ten, twenty, or even fifty years may elapse before a second flowering takes place. Indeed, there are species in general cultivation, such as *A. nitida*, that have never been seen to flower at all. In fact, the Chinese believe that *A. nitida* flowers only once in every century. At the other end of the scale *P. aurea* and *Sh. kumasaca* flower periodically from time to time with no apparent harm.

In the course of time a number of fallacious ideas concerning the flowering of bamboos have gained credence. It has been stated that when a stand of bamboo flowers in one part of a country all the other clumps of the same species flower also, not only in other parts of the same country, but all over the world as well. When *A. spathiflora* flowered simultaneously in 1876 in Europe, North Africa, and in Sikkim, and when *P. boryana* flowered in many parts of England, France, Germany, and Switzerland, some authorities, inspired no doubt by the phenomenon, stated that when a given species flowers in one locality it does so throughout the world. That statement is not only untrue, but is also botanically unsound. The simultaneous flowering in gardens is due largely to the relatively few introductions of bamboo species into Europe; so that the majority of our plants have come from dividing and redividing these clumps throughout the years. When clones are divided thus, each portion, however small, retains its original parental botanical identity. When, however, a plant is raised from seed its botanical make-up varies within narrow limits, and these variations give it the chance of evolving to suit altered conditions. The jungle bamboos spend most of their time growing vegetatively from clones, so if one or more clone did flower at the same time a great number of variable seedlings must arise from the vast amount of seed produced. The blowing pollen from the flowering bamboos gives the race the chance of altering itself to suit the changing environment. There do not appear to be any authentic records or available specimens of fossil bamboos. However, many species of other genera of grasses (the Gramineae) have been found in the Eocene and Oligocene beds of sixty million years ago.

The variable seedling theory was proved at Pitt White when over a period of fourteen years thirteen clumps of *A. fastuosa* were planted in the gardens. These were obtained from different nurseries, who in turn had acquired the original clumps from separate sources abroad. During the next thirteen years several of the clumps had flowered many times, others did so once or twice, then died, but four stands did not come into flower at all, and still thrive happily.

True hybridization between related species of bamboo has yet to be recorded, but in 1937 T. S. Venkatraman published an account of successful hybridization of *B. arundinacaea* with another type of giant grass, the Indian Sugar Cane. Before any progress can be made in this direction much more information must be acquired on such vital issues as the possibility of provoking premature flowering experimentally, and the viability of pollen stored for future service. Only then can plant breeding of desirable grasses be fruitfully pursued. It might eventually be possible to develop strains presenting one feature or another of commercial importance in optimal degree, such as an exceptionally long-fibred strain especially bred for paper-making, but at the moment this is no more than a dream awaiting future fulfilment.

The second misconception is that the bamboo always dies after flowering. This is a false assumption, and easily disproved, for many species, notably two mentioned already—*P. aurea* and *Sh. kumasaca*—do, so regularly without suffering serious harm. A partial flowering seldom causes the death of a plant, but a full, heavy blooming can have a shattering effect on a newly planted clump, especially if it happens to be one of those species that are shy cane growers. I have seen this happen to separate plants of *A. simonii* at Pitt White. I had taken several medium-sized divisions and some rooted basal cane cuttings from the well-established parent clump, and they all grew lustily, but during the following season they all suddenly burst into bloom. The flowering was extremely prolific, and the parent plant also flowered, but only on a few canes, and whilst the smaller plants threw witterings full of flowers, the parent plant continued to send up flower-free canes in the normal fashion. Having considered this fact of partial flowering, I have reached the conclusion that if the original clump is well established and is of a naturally vigorous type, it might possibly possess the necessary stamina to enable it to overcome the exhausting drain on its resources caused by the blooming. This is, of course, only my own opinion, but if the original plant manages to survive the flowerings, then it could prove my theory.

It used to be believed that the interval between successive flowerings was constant for a given species. Enough records are available at the moment to prove that such a generalization is not even approximately true. Authentic flowering records of two well-known species prove the point. *A. falconeri* flowered in 1876, 1890, 1929, and 1936. This means that this species came into flower at intervals of fourteen, thirty-nine, and seven years, in that order. *P. aurea* flowered in 1876, 1904, 1922, and 1936, thus showing gaps of twenty-eight, eighteen, and fourteen years between flowerings. It is obvious that the flowering habit does not form a regular pattern, and accurate prediction of a flowering period is not possible at the present moment.

Faulty identification of a species has caused a lot of trouble in the past, especially when the plant comes into flower. I know of one species, growing in a well-known garden, which was recorded as having been in flower, and I discovered to my astonishment that though it had been growing for many years under an Arundinaria label, it was, in fact, one of the Phyllostachys. That glaring error of identification is not as uncommon as one would think, and it goes to show that authentic identiacation of the species is essential, especially when flowering occurs.

What causes the bamboo to burst into flower? Opinions differ, and no one can say with accuracy. Weather conditions must be discounted, since the temperature ranges and humidity levels in different parts of the globe where the bamboo may be in flower are so diverse. It can be argued that the plant may have reached the limit of its growth cycle, and that the approach of inevitable senility and death urges it to begin the process of procreation quickly to enable the species to survive. Clearly there is some inborn time mechanism implanted when the seed is first formed. Its nature and mode of operation must remain a problem for closer botanical research.

In an effort to prolong the life of a flowering clump, some authorities believe that it is a good precaution to cut down all the existing canes in the hope that this would arrest the flowering, thus enabling the basal shoots to produce fresh canes which may be free from the flowering tendency. But I think that this operation should only be used as a last resort, as it could hasten the demise of the plant by terminating the food supply from the leaves to the roots. If the species concerned is of a type which makes only a few canes annually, the chances are that the plant will die before new canes are able to grow. Rather than risk losing an erstwhile valuable plant in this way, try to lessen the strain that the flowering imposes on the plant's resources by plucking out the spikes as soon as they appear, in much the same way as you would pluck the plumes from a stem of grass. I have found that doing this allows enough leaves to remain on the plant to continue their work of photosynthesis, thus building up the food store in the rhizomes and prolonging the life span of the plant.

It would appear that when flowering commences the rhizomes cease to make new branches and the latent buds on existing branches and on the base of the stem lose their power to burst into active growth. It is tempting to think that some hormone or chemical agent, manufactured in the flowering heads and transported down to the underground structures, inhibits these mechanisms. If this is indeed the case, it would be a logical procedure to cut off the flowering cane in the expectation that the basal buds, released from such inhibition, would send up new shoots. Unfortunately, these, too, may be dominated by the inherent impulse to flower. This, of course, is only speculation, and until more is known about the flowering mechanisms, it would appear to be much safer to remove the flowering heads by plucking as and when they appear.

Encouragement can, however, be taken from the experience with hardy bamboos over the last century. Most of the species, varieties, and cultivars imported into those Victorian and Edwardian gardens have flowered at one time or another during the period which has elapsed, yet all are still to be found in healthy growth in many parts of the country. The only types of hardy bamboo which probably stand in danger of becoming extinct are the cultivars, i.e. seedling sports from a species, which have multiplied from a single clonal source. Such a disaster so far remains unrecorded, but if it does happen at all it will probably be to the weaker cultivars.

⹋ 3 ⹌

Care and Cultivation

*Site and soil preparation—Planting out—Care of established
clumps—Winter protection—Pests and diseases*

SITE AND SOIL PREPARATION

Although bamboos thrive best in good open loamy soil, they can be
made to take root and grow freely in far less favourable media, provided
that the site has been well dug and manured in advance in the early
spring, and the soil has had time to warm up.

Drainage must be good. This cannot be stressed too strongly, as the
bamboo, which loves an abundance of water and a humid atmosphere,
will not grow in permanently waterlogged soil, and it is a waste of time
to try to make it do so. If the drainage is not all that it should be, it is well
worth while excavating several feet of the original soil and filling the
bottom of the hole with broken stones or brickbats, or drainage pipes if
necessary, to dispose of the surplus water. Bamboos may not like
standing with their feet permanently under water, but a position by
the side of a stream or on a bank above boggy ground will suit them
well. In a situation such as this they can obtain all the necessary moisture
for their well-being, without having to suffer from either a lack or
excess of it. On the other hand, bamboos will not tolerate arid conditions, and unless adequate moisture is available the shoots will dehy-
drate quickly, and even those clumps which are well established in a
site previously moisture laden will not thrive, but grow stunted canes
and lack-lustre foliage.

If the planting site is already one of high fertility, all that needs doing
is to add a little humus for enrichment, thus enabling the young plants
to start quickly into growth. The type of humus which gives the longest-
lasting supply of food is good farmyard cow manure. But any kind of
manure, with the exception of fresh pig manure, will do, provided that
it is well rotted and short in texture; and though fresh pig manure
is unsuitable, due to its high acid content, pig manure can be used

41

provided it has been left to mature in a heap for at least a whole season. Trench the manure in deeply, mixing it thoroughly into the soil. To provide additional food later on when the nourishment in the manure has become exhausted, add several handfuls of coarse bonemeal, the coarser the better, into the original mixing.

If, however, the site contains poor-quality soil, it may be advisable to remove the original material and replace with fresh soil of a better quality, to ensure that the young bamboos have enough food to last them until they are able to send out their roots further afield in search of sustenance. The excavated area need not be large, a space of 4 feet square and 2 feet deep being quite sufficient. As advised for the good soil site, add a liberal dressing of well-rotted short manure, plus an even heavier dressing of bone-meal. The soil preparation of the planting site may seem to entail a lot of extra work, but it must be emphasized that this work is essential. Skimping the preparation, the good loam, or the manure is simply not worth while, because together these add up to the vital nourishment needed to make the roots and rhizomes grow quickly, particularly during the first critical years.

If the soil in your garden is poor and impoverished, do not be despondent, for the hardy bamboo will tolerate many types of soil, acid and lime-laden alike, the only essential for their establishment being that the site should be well prepared in advance of their primary planting.

PLANTING OUT

Geographical factors and local weather conditions determine the propitious moment for planting out, and should be given careful attention. In Europe the weather varies from place to place, but certain generalizations can be made, at least as far as the British Isles are concerned. Broadly speaking, the warmth of the local climate in the United Kingdom can be gauged by the amount of electric current required to maintain a stated level of temperature in a standard greenhouse. Relevant data published by the Electricity Board on this subject indicate that no matter whether the requirement of the whole winter season be considered, or simply those of the early spring, the United Kingdom can be classified into several zones:

Devon, Cornwall, and Wales enjoy the warmest winters and the earliest springs. Dorset and Hampshire are a little cooler. The Home Counties fall into the next group. The Shires and Midlands are cooler still, whilst Scotland and the northernmost counties of England are the coldest places, although showing a strong tendency towards milder conditions in their western halves. The position is, however, modified greatly by warmth from ocean currents associated with the Gulf Stream, which keeps certain coastal belts much warmer than near-by inland areas. This effect is greatest along the coast of south-west England, the south and west coasts of Eire, in the Scilly and Channel Isles, and in the fringes of the western highland coastal areas of Scotland. With such wide variations in climate and in the dates of the horticulturist's spring, it cannot be emphasized too strongly that each prospective planter of hardy bamboos must study his own local weather conditions and prospects with great care, and always defer planting out bamboos bought from any nursery until the local conditions are suitable.

The propitious date varies from early spring in the warm Gulf Stream areas to late May in other parts of the country, or it may even be early June where the spring comes late.

It is folly to plant any species of bamboo whilst the soil is cold and unreceptive. The probability is that the roots and rhizomes will shrivel away and the reproductive buds rot. The roots and rhizomes will not in any case start to move until the soil is warm enough, and premature planting into cold damp soil can even cause defoliation of the existing canes in the new clump, particularly in the taller-caned types. Plants obtained from a nursery can always be stored for a period if they are kept moist and protected, so if you are in doubt store the plant until weather conditions are more suitable.

There is another question to be considered. At what stage in the cycle of growth is the bamboo in the right condition for lifting and transporting? The most appropriate moment would appear to be soon after the underground buds, from which the new season's crop of canes will emerge, have become active. If the bamboos are purchased from a nursery with seasonal conditions approximating your own, the ideal lifting and planting time will be similar, and the plants can be accepted confidently as soon as your site begins to feel the full warmth of the spring air and sun. If, however, the nursery enjoys a much milder climate than your garden does, its stock will become active before the soil in your garden site is ready to receive them. There are two courses open to you:

1. You may ask the nursery to delay delivery of the bamboos pending your notification that you are ready to plant, but by then the tender new shoots may be well above ground and the plant as a whole more vulnerable to the inevitable hazards of the journey. (Bamboos in this

condition growing in your own garden, however, will transplant readily from one position to another quite safely.)

2. Accept delivery of the bamboos at the nursery's time, and plant them as soon as they arrive in a large pot or box, covering the whole plant with moist soil or peat, and store in a warm shed out of the cold, until the time is ripe for planting out in their permanent sites. Bamboos stored thus will live for many weeks, and as long as the covering soil is kept moist, and the whole free from cold, there need be no fear of losing them. There will always be the inevitable know-all who will tell you that he can plant bamboos at any time of the year in his garden. He may be lucky, or more probably he is a liar, and it is wise to ignore him. Better by far to have a healthy bamboo growing temporarily in a box than have a costly bamboo sitting in the garden dead and blasted. My advice is based on long experience, so do not plant hastily, and never before conditions are right.

The more commonplace species of bamboo sent out by the nurseries are usually either parts of a clump taken by division from an established clump or it may be one of the re-divided smaller plants. The rare or more uncommon species may be sent out as rooted rhizome cuttings or basal cane cuttings, usually pot raised. The larger division will naturally provide a faster-growing new clump, but the cuttings do, in the course of time, make just as good plants.

The chunks taken from the established plant should consist of a close-knit mass of tangled roots and rhizomes, and some mature canes, plus a few fresh shoots and some dormant buds. The pot-raised plants usually consist of several short new canes, but minus the web of older active roots. If the potted plant is on the small side, it is wise to repot it as soon as it arrives and keep it in a greenhouse until more vigorous canes begin to appear.

You may find when you receive your plant, if it is a large portion from an established clump, that the grower has cut down the mature canes to a few feet high. This is done solely to facilitate packaging and posting. It does no harm to the clump itself, as it is the new shoots that matter most of all. As a matter of fact, I prefer to cut down the older canes to a few feet when transplanting any clump of vigorous-growing bamboo, as this seems to encourage the semi-dormant shoots to grow more quickly.

Whether you receive the large divided portion or the smaller potted plant, take great care when unpacking it, as the young spears, which are the new canes of the coming year, are extremely fragile; so fragile are they, indeed, that they will snap off easily with careless handling. Always remember that the eventual establishing of the new clump depends entirely upon how these delicate shoots are handled at the outset, so be extra careful.

If the planting site is not ready to receive the bamboos right away, or if the weather is inclement, they can, as I have said before, be stored in a warm spot out of the wind. In any case, whether the site is ready or not, it is advisable to soak the rhizomes and roots overnight in tepid water. This counteracts any drying out which may have occurred during transportation.

When planting, the general rule should be to adopt the level prescribed by the original soil markings on the plant. This is easily recognizable at the base of the canes, where the colour of the stems begins to fade from green to creamy white—the colour of the rhizomes attached to it. The general system of interlocking rhizomes will then be at distances from 3 to 4 inches, or up to a foot from the surface, whilst the short roots projecting from them will penetrate more deeply.

Before planting, fork the bottom of the plot lightly, and place the plant in carefully. Sift the soil through the roots and rhizomes to fill any air spaces, and press down firmly. When the site has been filled to the level of the surrounding ground, tread around the clump delicately, being careful to avoid the new shoots, some of which may be poking their points through the soil surface. Finally saturate the area thoroughly with rain-water. Keep the soil in the immediate vicinity free from weeds and grass, since grass, especially the taller-growing kinds, can smother the young canes before they can rise above it. Given relative freedom from competition in their youth, the young bamboos will grow rapidly, and as they branch out and unfold their mantle of foliage any weeds growing in the vicinity will quickly be dominated.

Nothing else need be done during the first season, other than to provide a copious supply of water during a dry spell. Once the fresh canes have begun to grow the soil around the roots should never be allowed to dry out. Even a short period of drought may severely damage and even kill a new plant. Constant superivision is essential until the growing period ends in the late summer. I think it is relevant at this point to mention that a note received from A. M. Kormilitzin, Curator at the State Nikita Botanic Gardens at Yalta in the Crimea, tells me that all the bamboos under cultivation in the gardens there require irrigation, and that they also suffer from air drought, but species growing in the more humid subtropical region of the Caucasian Black Sea area

(Sochi, Sukhumi, and Batumi) do not require irrigation and are free from air drought. In the United Kingdom there is little to fear from air drought, but care must be given to see that the moisture supply to the roots is not lacking.

CARE OF ESTABLISHED CLUMPS

Hardy bamboos can be divided into two sections. These are known as 'caespitose' and 'running' types. The 'caespitose' types do not send out long rhizomes and tend to remain in a close group; they may be well furnished with new shoots every year, but do not stray far from the place where they were originally planted. These are obviously the type of bamboo to use as specimen or ornamental plants, or for planting in a restricted position. The 'running' types may sometimes be rampant in their mode of growth and often need keeping in check, but in most cases control can be maintained quite simply. The dwarf species can be cut back with a spade, since the rhizomes are thin, and those with thicker tougher rhizomes can be controlled by breaking off the new shoots as soon as they appear above ground. After this treatment further new shoots may appear later on in the season, but usually at a shorter distance from the parent clump. These, too, can be snapped off easily if found outside the growing area. Rotovating the ground around the clump, or, when the surrounding grass is short, mowing regularly around the perimeter, will restrict the standing growth to a prescribed zone. It is surprising just how far away from the parent plant some of the shoots can appear. I have observed shoots of *B. quadrangularis* emerge from the ground 12 feet away, in a direct line, from the nearest previous year's clump of canes.

It is important therefore, when selecting a hardy bamboo, to choose one appropriate to the site which is available. Unless the ground is poor, and the conditions harsh, a vigorously 'running' type is unlikely to make a plant of circumscribed dimensions.

Each clump must be given room for expansion. A stand of bamboo usually continues to grow and increase in size for many years; in fact, from one human generation to the next. If the plantings are made at intervals of less than 12 feet apart, the groups will eventually meet and grow together until their distinction is lost in a thicket of canes.

All established clumps ought to have a top dressing occasionally. Where the surrounding soil is rich, a dressing every other year should suffice, but where the soil is of a poorer quality, an annual dressing is advisable. The top dressing can consist either of good, well-rotted manure or compost. Well-decomposed leaf-mould can be used, but should be fortified with additional food such as fish manure, or any similar material which has a high nitrogenous content. Spread the dressing generously on the surface of the soil, but do not attempt to dig it in. Disturbing the soil surface with forking could damage the rhizome growth, and could snap off new shoots below the surface. The mulching should therefore be confined to packing the dressing around the perimeter of the clump, and if the canes are not too tightly packed, apply some there also.

Early autumn is the best time to apply the dressing, as this allows the winter rains and frost to break it down, and wash the nutriments into the soil. Grass clippings, which are often hard to dispose of, can be used as a supplementary dressing, but not as a main mulch. It should not be heaped on too thickly, as grass, especially when wet, can generate enough heat to cause damage to the soft cane growth.

Simple plant hygiene is important when looking after an old-established bed of hardy bamboos. All dead or partially expiring canes should be removed when seen, and so should any canes which are broken or distorted. Nothing spoils the appearance of a symmetrical green group of bamboos more than a few gaunt dead canes projecting their bare bones through the top of the clump. In any case, these dead or damaged canes not only look unsightly, but they provide a breeding-ground for pests and fungi. Dead leaves lodged between the canes should be shaken free with a hooked stick. There is no need to pull each leaf out; let them fall to the ground inside the clump, where they will decompose quickly.

When canes are required for garden use always use the older canes first. These are easily recognized by their fading, rather sparse foliage, the die-back from the tip, and their general buff colour. Use a large-diameter branch cutter to sever the bigger canes, and never attempt to break a cane with the bare hands; few of them snap easily anyway, and those which split can cut the hand like a razor blade if handled with less than extra care.

WINTER PROTECTION

Well-established hardy bamboos need little in the way of winter protection unless they are planted in very exposed positions, and only then if they are one of the more delicate species. Excluding the semi-tender species, the rest are much more hardy than most people realize. In the gardens at Pitt White, where there are over fifty species of hardy bamboos growing out of doors, not one clump was lost during the severe winter of 1962/3, when the temperatures sank to their lowest for nearly a century. It is true that some species may lose a portion of their foliage during a prolonged hard winter, especially if they are growing in a very exposed position, but fresh leaves will appear again on the denuded canes in the following spring. In exceptionally cold weather the canes of some species may themselves be cut to the ground, but the rhizomes are seldom if ever damaged.

If it is felt, however, that a large clump does require protection, any woven hurdle or split chestnut fencing secured to the windward side of the clump should provide all that is needed, but it must be understood that some species of bamboo grow very tall canes indeed, thus making the provision of artificial shelter difficult.

All new plantings need some kind of frost protection, no matter what species is being cultivated; even the hardy types may suffer if needlessly exposed to frost and snow, as well as from the chilling winds and rain. Young bamboos planted in a sheltered spot in the appropriate season—during the spring or early summer—should not require protection until the autumn, but those planted in a windswept site *must* be given shelter as soon as they are in the ground. Any type of temporary shelter is sufficient to last until the autumn; then something more substantial will be needed.

There are several means of providing shelter. It can be done by covering over the plant with sacking, cloth, or polythene. Sacking is good as long as it can be kept fairly dry, but it tends to promote the growth of mildew, and polythene is too light and is difficult to secure. After using many different types of material, I have found that the most successful is the 'house' method.

These 'houses' are easily constructed, are not unsightly, and cost little or nothing in the way of material. To build them, place a ring of canes at a few intervals apart around the clump, keeping each cane 2 or 3 inches away from the outermost-growing canes. Then make another ring, leaving a 2-inch gap between it and the inner circle. Tie the tops of the inner ring of canes together, making sure that the top of the tallest cane in the growing clump has at least 2 inches of headroom. Pack dry fern or bracken between the rings, working around the circle and packing down firmly as you go, until the packed fern reaches just over half-way to the top of the clump. Tie the tops of the outer ring of canes to the inner, and you will see that the basic roof pattern has been formed. To clothe this, tie bunches of dry fern to the top of the 'house',

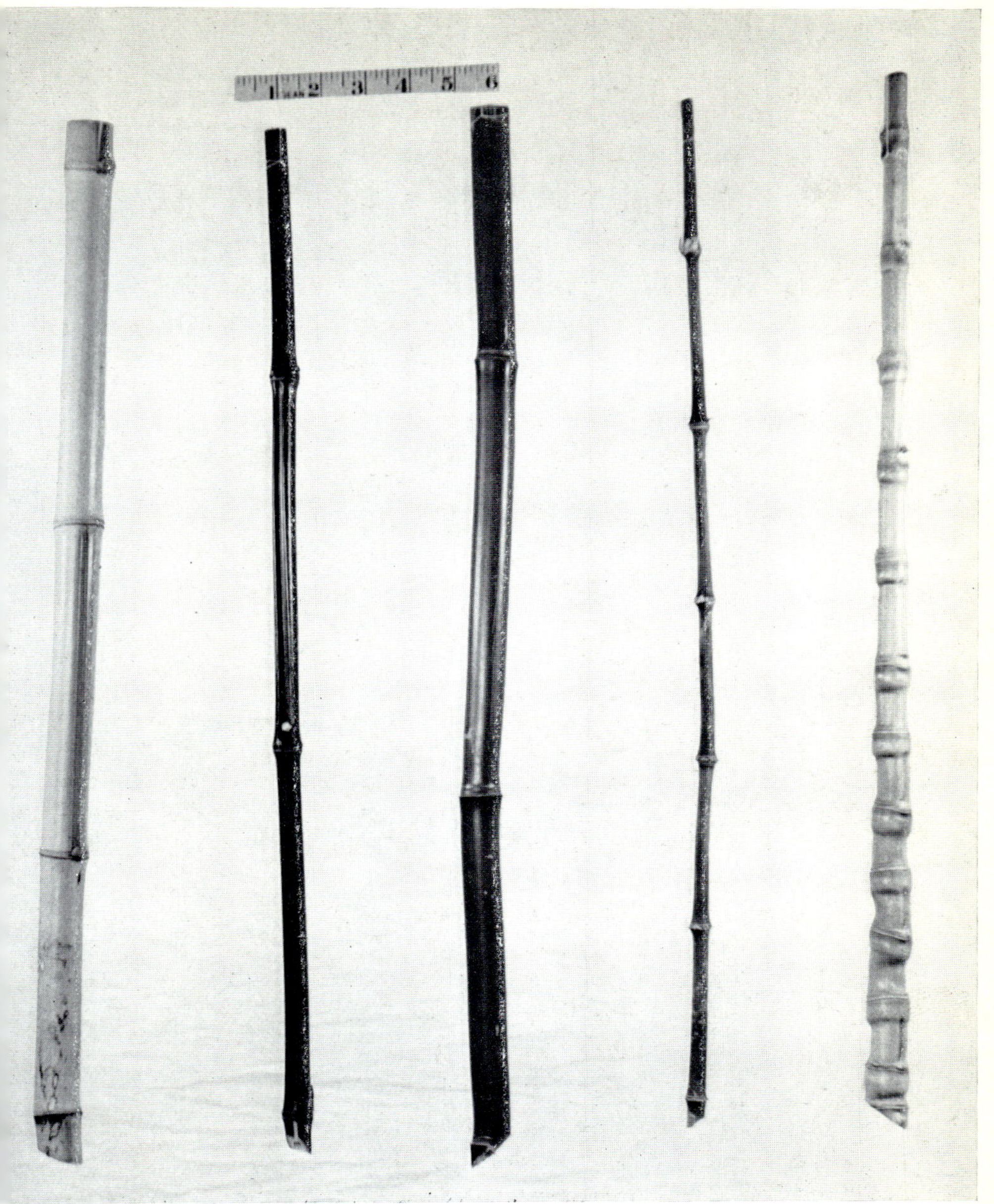

5. Different types of cane structure. *P. castillonis*—*P. nigra*—*A. fastuosa*—*B. quadrangularis*—*P. aurea*

6. Flowering sprays of hardy bamboo, Pitt White, 1963: (left) *P. castillonis* (right) *A. fastuosa*

7.
Flowering sprays of hardy bamboo, Pitt White, 1964: (left) *A. hindsii* (right) *S. palmata* var. *nebulosa*

letting the free ends of the fronds hang downwards. Continue doing this until the whole roof has been thickly and evenly clothed. Finally bind the whole erection with a few turns of twine to secure the loose ends, and the 'house' is tailored to fit the plant. This simple construction breaks up the direct force of the wind, sheds the rain and snow, but does not interfere with the free circulation of air around the protected plant.

If properly made, one of these structures can last for a whole winter. During the milder spells which sometimes occur during the colder months a gap can be made in the roof or in the side of the 'house' to allow more air to circulate, and for sunlight to reach the leaves. The gap can quickly be closed when the weather again becomes inclement. The foliage does not suffer in any way by being deprived of direct light when this method is used. The green colour may fade somewhat, but the rich green soon returns when the shelter is removed in the spring. I have used this method for many years without disappointment. During the bitter winter of 1962/3 I used it to shelter a new planting of *P. mitis*, well known to be a difficult subject to get started in this country, and the plant survived the winter without suffering more damage than a little marginal leaf scorch.

As a substitute for fern or bracken, dry leaves can be used to fill the sides of the 'house', but wire-netting will then be needed to hold the walls together. Hay is not much use, but straw, if it is long enough, can be useful. But without doubt fern and bracken make the warmest jacket, and in the spring it can be chopped up for use as a top dressing on the surrounding soil. Never heap leaf litter or grass over a young planting and expect this to give protection during the winter. Not only is this hard to confine in a heap, but it rots down too quickly, and simply provides a good shelter for slugs and pests of that ilk.

PESTS AND DISEASES

Hardy bamboos grown in temperate climates are singularly free from damage by microbial and insect pests, and it is only amongst the young canes that damage is likely to occur. Simple garden hygiene is usually sufficient to forestall damage. Even in the warmer climates, plant diseases do the really hardy bamboo species comparatively little harm.

Pests

Slugs and snails sometimes like to chew on the soft young spears just when they are breaking through the surface of the soil. It is during this brief period that the shoots are at the mercy of such predators. Bamboo seedlings are especially vulnerable, and when tiny can be totally destroyed by slugs in a single night. The danger can be met effectively by scattering metaldehyde bait around the planting area, and by keeping the immediate surroundings free from leaves and general debris which is likely to harbour such predators. Mice occasionally like to nibble the tender new shoots, especially during the early spring, and they have been known to strip the sheaths from the canes to line their nests. The rodent killer Warfarin easily disposes of these pests. Convenient mixtures of this substance in the form of bait material are sold under various brand names. The bait must be protected from the rain, and as it is a poison must be made inaccessible to children and household pets. This can be done by placing the bait in a small cylindrical drainpipe arranged horizontally, or under a tile or slate supported on small stones, and held in position by a larger stone placed on top.

Cats and dogs, too, can cause much damage by selecting the newly turned soft soil on the planting site for toilet purposes, or for burying bones. It is then that the new shoots may be broken off inadvertently. As a preventative but temporary measure the site can be strewn with chopped gorse or thorns, and sprinkled with repellent powder until the canes are large enough to do without it.

Some birds, starlings in particular, love to congregate in thick clumps of bamboo towards dusk, and cause offence with their constant chattering and rustling. Their droppings may smear the canes and make them look unsightly, but the birds themselves seldom damage the canes and cannot seriously be classed as pests. The same applies to blackbirds and thrushes, who often build their nests in a close-growing thicket.

Naturally there is always the possibility of casual damage by larger animals. Livestock greedily consume the leaves and young shoots of the bamboo, and adequate fencing must be provided around any plantings if cattle, etc., are allowed to wander in a grove. Smaller animals, such as rabbits or badgers, too, can cause damage by blundering through the clump and breaking the young canes, but these hazards are always with us and each gardener must guard his own plot.

Leaf-eating insects (earwigs, thrips, etc.) have been known to spoil the foliage of the younger canes before the leaves unfold. Earwigs occasionally eat the tips of the soft leaves, or use them as places of concealment, but in my experience the damage caused by them has never been more than trivial. In the subtropical regions attacks by various types of aphis have been reported, but these have quickly been brought under control by the use of Malathion. As the foliage ages and progressively toughens damage by these pests lessens and their effect becomes negligible.

Diseases

Once the bamboo has attained the status of an established clump it is relatively disease free. It is true that some species do suffer from withering and decaying of the leaf margins and tips; the Sasa are singularly unlucky in this respect, as practically every member suffers to some extent from this peculiarity. It is usually considered to be an inherent tendency rather than an acquired disease, as it affects these bamboos in their native habitat in just the same way as it does in our gardens. Up to the moment of writing the disease, if it is one, has resisted all attempts at prevention. Anyway, at a little distance this marginal withering suggests variegation, adding interest to those species which have an otherwise drab appearance.

Bamboo smut is a recognized bamboo disease in Japan, and it is also known in the tropical areas of India and Burma, but up to date there do not appear to be any recordings of its having been active in Europe or the U.S.S.R. It has cropped up on several occasions in various parts of the United States of America and was eradicated when intensive counter-measures were taken. The action was drastic but essential, as every bit of the affected plant, including the rhizomes, had to be rooted out and incinerated.

Another disease, known as *Melanconium culm*, has been reported in Great Britain and in several parts of the continent of Europe. I myself have seen this disease at first hand on one bamboo (*P. castillonis*) in the gardens at Pitt White, but the damage caused was slight and easily brought under control. It seems only to attack canes which have suffered damage already, the primary focus being near the base of the cane. The internode above the part attacked turns black, and the disease travels up the cane until the entire cane is involved. The remedy is to cut out the cane completely, including any damaged portion of the rhizome. Cut back until healthy wood is reached, then dust the cut and the surrounding area with green sulphur powder. No one knows how this disease originated, or indeed how many of the other diseases start, since comparatively little research has been carried out on the diseases of the bamboo, and until more has been done reliance must be placed on the cleanliness of the original stock and on good husbandry to keep pests and diseases away.

Few diseases attack bamboos growing in the temperate zones, and in the countries of western Europe they are of little practical concern. From what has been said previously it should be clear that the hardy bamboos as a whole are usually accommodating plants, being more rubustly healthy than most other garden plants. Of course, it is true that they require special attention when first planted, and that small nursery plantings and the more tender species need attentive supervision throughout the first two vital growing seasons, but eventually the established clump will take care of itself, and be an object of beauty and a source of pleasure to you and your family for many years, and indeed for decades to come.

The Propagation of the Hardy Bamboo

Clump division—Rhizome cuttings—Basal cane cuttings—Cane layering—Nodal cuttings—Growing from seed

Bamboos are propagated mostly by vegetative means, as obtaining seed from the flowering is not an annual occurrence, but only repeated at intervals of many years, and few species introduced into temperate climates from warmer places set fertile seed. There are five methods of vegetative propagation, ranging in degrees of simplicity and effort involved, from the easiest, which is also the most commonly used, i.e. clump division, to the new and little-tried method of taking nodal cuttings. Each of the methods described, with the exception of the nodal cuttings technique, can be relied on to provide ample fresh material at comparatively little cost. Propagation from seed has also been described, in case you are lucky enough to obtain some.

CLUMP DIVISION

Division of the clump should be carried out in the spring, when the new growth is stirring. This is the best time in most areas, but in very mild districts clumps can safely be divided at any time of the year, with the exception of the coldest winter months. If, however, the new plant is to be transferred to a colder site, it is always advisable to carry out the operation in the spring, with the whole growing season ahead in which it can accommodate itself to its new quarters.

Propagation by division is a long-established practice, and can be done crudely with a spade, but when more than one plant is needed it is far better to lift the clump, lay it flat upon the ground, and then divide it into suitable parts which can be dissected off without causing undue damage. A greater number of divisions can be secured in this way, and these, if of sufficient vigour, can be transferred to permanent sites. The smaller ones should always be placed for a period in a nursery bed until they show clear signs of having formed new roots and rhizomes, and are sending up new shoots.

If a large number of plants is required, small nursery plants can be re-divided during the season which follows. When this process of sub-division has been continued for some time, the whole feltwork of roots and rhizomes becomes dwarfed, and produces numbers of slender shoots instead of the long typical canes. From such material even very small pieces will suffice for the multiplication of stock, if warmth, combined if possible with mist propagation, is provided. The application of this method is perhaps of greatest value when the rarer species of bamboo are being considered.

The small plants raised by repeated subdivision may bear little resemblance to their elders which are depicted or described in the nursery catalogues, but once established in well-prepared permanent sites quickly develop larger rhizomes and shoots and regain their true characteristic appearance. They have the special advantage of having an individual rooting system free from the dead and damaged material found in the clumps which have been crudely cut off with a spade. Their comparatively low weight is an added factor, and a favourable one when transport charges are important.

No matter which scheme is adopted, all the work of dividing up the plant should be done with a sharp pair of suitable secateurs, and precautions taken against dehydration caused by exposure to wind and sun.

The lifting of large stands of bamboo presents its own mechanical problems. A tripod with block and tackle may be needed, but the root system of the hardy types in this country does not extend far into the ground, and removal can be facilitated by surrounding the clump with a deep trench and exercising a levering action with a spade driven inwards under the mass.

In warmer climates, as in the United States Research Station at Puerto Rico, where they grow the massive tropical species, the solid interlocking mass of thick and ivory-hard rhizomes can be very difficult to deal with indeed. It is recorded that on one occasion a monster clump was split up by the simultaneous detonating of several sticks of dynamite inserted in holes bored obliquely around the perimeter of the clump. It was laconically reported that the blast loosened the clump sufficiently and facilitated division. . . . I bet it did. Fortunately there is little danger that such extreme measures will ever be needed in the temperate countries.

RHIZOME CUTTINGS

Propagation by rhizome cuttings is a much less physically exhausting process which can be used when dealing with the 'running' species of bamboo, and particularly with those whose rhizomes stray far from the parent plant, and therefore can be separated easily. The cuttings should be obtained and planted early in the season, before shoot development has begun in earnest. Most types of hardy bamboo grown in the United Kingdom, and especially most of the Arundinaria, are of the 'running' type, and carry embryo cane buds on nearly every rhizome joint. These are the species which provide the best and most reproductive material. There are a few exceptions to this rule, notably the unusual *A. niitakayamensis* (Pitt White clone), which sends out long connecting links furnished with embryonic cane buds only at the tips. The rest of the rhizome from tip back to the starting-point is of no value whatsoever for propagation purposes.

On most types of 'running' bamboo, however, rhizome cuttings are obtained quite easily. If a previous year's rhizome is pulled free from the ground, it will be seen that it leads back to the closest cane of the last year's growth. The whole rhizome can be cut into portions, each containing several joints. On the majority of these joints there appear semi-dormant growth buds. Fresh young rhizomes must be used, those more than two years old being of little propagative value.

The cut portions can be placed in a trench, previously dug some 6 inches in depth, and planted about 10 inches apart in a vertical position. The tip of the rhizome should just show above the ground-level, and the topmost bud will quickly elongate to form a cane. (The vertical and horizontal planting positions are explained in the section of Chapter 1 dealing with hedges.) In the following spring the rooted cuttings can be bedded out a few feet apart in a nursery plot, and transferred to their permanent sites, if these are not too rough, a few years later on.

Smaller quantities of rhizome cuttings can be propagated in pots or boxes containing ordinary garden soil. Planted thus in the spring, they require no shelter until the autumn, when the pots or boxes should be brought indoors to a warm greenhouse. Given this winter treatment, the cuttings will continue to grow steadily, sending out fresh shoots and roots at a time when the growth of their brothers planted out of doors is at a standstill. Within two seasons a rhizome cutting can, if given this help, be large enough for planting in its final position as a viable clump.

Cuttings planted in pots, boxes, or trenches must never be allowed to dry out. A daily syringe with tepid water, especially for cuttings in a warm house, helps to keep the air humid, a very important condition for the healthy growth of all bamboos.

There seems to be little gained by using hormone rooting powder as an additional aid to faster rooting. My own trials, using three different strengths of hormone powder, showed that the rhizome cuttings treated thus did not root any faster than those inserted free from powder.

BASAL CANE CUTTINGS

This is a method suitable for all types of bamboo, but particularly suitable to the 'caespitose' types, which grow in circumscribed clumps and do not form long wandering rhizomes.

Second-year cuttings selected from the perimeter of the clump are the best for this purpose. Pull the soil away from between the clump and the cane to expose the connecting rhizome. This is severed and the cane lifted with as much soil left attached to the roots as possible. Trim the exposed cuts of the roots and rhizome and dust these with green sulphur powder; this is a precaution aimed at minimizing the effects of bacterial damage. Cut the cane down to within a foot or so of the base, leaving if possible at least one or two internodes intact. Plant the cutting in a deep pot or box containing sieved ordinary garden soil. The plant receptacle must be of sufficient depth to allow the cutting to be inserted below the original soil level, so that the lowest node is well below the surface. If there are several nodes crowded closely together at the base, as in the case of *P. aurea*, they can all be buried. Dormant shoots are well represented at the lower nodes of the canes in most species. In close association with them are the dormant or semi-dormant verticillate roots which can be seen as a ring of embryonic granules, or as actual pendulous aerial roots, either healthy or desiccated, hanging downwards towards the soil, although barely reaching it. If these areas are buried in warm soil, the dormant buds become active, and the cutting quickly roots and new spears begin to shoot.

Basal cane cuttings can be taken at any time of the year if one has a warm house in which to grow them, but if warm house shelter is not available the best time is late spring. Fresh shoots should appear within two months of taking the cutting, and if left in the warm house they will continue to grow throughout the winter months. The cutting should be left to grow undisturbed for two seasons before planting out in a permanent garden site. In the second the pot or box should then be plunged in a bed of ashes or peat, and the cutting will by then be strong enough to stand the winter out of doors.

This is a well-tried method, almost always giving satisfactory results. It is simpler in operation than that of clump division, and provides a plant large enough for the permanent site more quickly than can be provided from simple rhizome cuttings.

CANE LAYERING

This method is not applicable to all species of hardy bamboo, but I have had some success with *B. quadrangularis*.

At the Ministry of Agriculture Research Station at Rosewarne in Cornwall, F. W. Shepherd used this method to produce a long hedge of *A. japonica* which still flourishes. The results of his trials show that new plantlets could be reproduced by the use of layering. Four types of material were used in the trials, from one- and two-year-old canes, some with roots attached and some without. All were buried in a horizontal position in 6-inch-deep trenches. After two years of undisturbed growth all four types of material had produced a number of new plants, but it was the two-year-old rooted canes which gave the best results. In fact, it was stated that on these canes new shoots arose from practically every node. It must be recorded with all fairness that subsequent attempts to achieve the same results were singularly unsuccessful.

The American botanists Young and Haun also describe the use of this method in the United States Department of Agriculture handbook on bamboos. Their technique is a modified version of the previous one, and they used it to propagate the tropical Bambusa and Dendrocalamus species.

In other hands and places success has not been so complete, and there is a tendency for the new shoots to arise only from the tip and the basal nodes of the layered canes. Perhaps this method is more suitable for trial in warm and humid localities than in the less-favoured areas. Its great attraction is undoubtedly that if successful it will provide a hedge or screen for little expenditure of either time or money.

NODAL CUTTINGS

The last vegetative method of propagation, from nodal cuttings, is reported to have been employed in China and Japan for many years. It requires material from thick-walled and large-diameter canes if it is to succeed. In hot climates these thick-diameter cuttings are easily come by, but are not so easily obtainable in cooler countries.

The cuttings should be taken from mature canes by sawing through half-way between the nodes. The slender topmost part of the cane and the lower basal extremity should be discarded. Each cutting must bear at least one short leaf-bearing branch. If there is more than one branch on the cutting, discard all those except the ones which face in the same direction. The open ends of the cutting are then packed with damp sand or soil, preferably sterilized. The cutting is then placed at twice its depth in a large pot containing John Innes No. 2 Potting Compost or a similar sterilized compound. The branch(es) should protrude above the soil-level, and lean at an angle of forty-five degrees to one side of the pot. If the cutting is kept in a close and warm environment and sprayed with tepid water every day, rooting should begin within two to three months.

Having made several attempts at this method, some with the assistance of hormone powder, and using many species, I have had a successful rooting only once. The successful cutting was taken from *B. quadrangularis*, and since then I have discovered that nodal cuttings root freely when taken from that species. But one must disregard that one species, and I have come to the conclusion that this method is not to be recommended to the amateur, and should be left to those who have ample facilities and an abundant supply of material which they can afford to lose if necessary.

GROWING FROM SEED

Any particular species of hardy bamboo is unlikely to flower more than once in the course of many years, and, as I have stated in a previous chapter, occasionally they may do so simultaneously in a number of countries. It does not follow, however, that when a bamboo does flower in the British Isles it will set seed; in fact, it rarely does so.

Bamboo seedlings raised in this country usually come from imported seed, and although I have had the fortunate experience of being able to raise seed from English-grown plants, the opportunities were extremely fortuitous. During the 1964/5 seasons *P. bambusoides* var. *castillonis* bloomed in many gardens all over the United Kingdom, but as far as I can ascertain the clump at Pitt White was the only one to produce viable seed. It was a fortunate circumstance that the peak of the flowering occurred during an exceptionally hot and sunny summer, with good weather extending long into the autumn. From that first flowering I was able to obtain some hundreds of sound seed, many of which germinated when sown. But in the second season, when the flowering was even more prolific, I was only able to get a few dozen seed from the hundreds of spathes of flowers when the canes were cut down. During that same season I obtained *two* seeds from a flowering clump of *A. graminea*, one of which germinated and is now a healthy sturdy plant. At the moment of writing *A. simonii* and *A. falconeri* are in flower, and swarms of sparrows and finches are busily searching for the ripening seed. In order to be able to collect some ripe seed for sowing I have enclosed several sprays of flowers in the old-fashioned peach ripening bags. They were never able to save my peaches but they are good at keeping birds at bay and the tiny ventilation holes let in enough light and air to assist ripening.

From a private communication from Mr. N. D. Treseder, who in his gardens at Truro has many interesting bamboos collected from historic gardens or rescued from obsolescent Victorian and Edwardian gardens in Cornwall, I have learnt that *A. falconeri* yielded plenty of fertile seed in his county when it flowered in 1935, and again in 1962.

These instances show the difficulty there is in obtaining seed from the home-grown bamboo, but if the reader has been fortunate enough to obtain some, the requirements for growing them are as follows.

Firstly, the seed should be sown as soon as it is ripe, as the home-produced seed soon shrivels and loses its virility. Secondly, sow the seed in sterilized soil and keep the pots in a good warm environment. I have experimented with several seed-sowing mixtures, ranging from pure sand, pure peat, ordinary garden soil, to the well-tried J.I. potting mixtures. The outcome of my trials showed that the ideal media was the J.I.P.2, lightened with double the quantity of silver sand. This mixture contains the correctly balanced nutriments for the growing seedling, and has the essential quality of having been made up from sterilized soil. If you cannot obtain J.I.P.2 in your local shop, or if you wish to make up your own mixture, the recipe is:

Parts by bulk: 7 sterilized loam
            3 peat
            3 sand (this is 1 part more than is used in the official
            J.I.P.2)
     plus $\frac{1}{4}$ lb J.I. base per bushel
          $\frac{3}{4}$ oz chalk per bushel

The chalk used can be either ground chalk, limestone, whiting or lime-stone flour.

The J.I. base itself can be made up by the gardener. The formula is:

Parts by weight: 2 hoof and horn, ⅛ grist (13 per cent nitrogen)
     2 superphosphate of lime (16 per cent phosphoric acid)
     1 sulphate of potash (48 per cent pure potash)

Analysis: Nitrogen 5.1 per cent, phosphoric acid 6.4 per cent, potash 9.7 per cent.

Sift the soil through a ⅜-inch sieve before sterilizing, and if the peat is coarse sift it through the same mesh sieve. Use dry sand in the mixture, and when the whole is thoroughly mixed you will find that the compost will hold together without packing or compressing.

The bamboo seeds, which vary greatly in size and shape depending upon the species, are known botanically as grooved caryopses in general form, and resemble their distant cousins, wheat, rice, and barley.

The seed should be sown flat, at a depth approximately equal to twice its own thickness, in pans of the compost. These should be placed in a warm and humid corner of the greenhouse and shaded from the direct rays of the sun. A seventy-degree propagation unit is eminently suitable for the purpose. Germination should take place within three to four weeks if the seed is sown as soon as it is ripe, and in three to twelve weeks when sown later in the year.

The first tiny spear will only produce a few small sheath blades at first, and remain static in growth for several weeks. As the daylight lengthens the growth will proceed apace, and the early shoot, which at first resembled a sprouting grain of wheat, soon begins to look like a miniature cane, bearing a succession of minute shoots, each tipped by a green blade; the true leaf at this stage. Extra shoots then come through the soil immediately around the original one, and grow larger as the rhizomatous root systems expand. When three or more shoots have appeared to reach their full length, the seedling should be lifted from the pan and potted into a two-inch pot containing the same type of compost as before. Once the root system has begun in earnest the small canes start to branch, but usually this does not happen until the second season of growth. With the branches come the true leaves, which, although small, are morphologically identical with those on the canes of a fully developed clump. By this time the seedling has become a recognizable bamboo plant of possibly a foot or more in height, and has up to a dozen small but well-leafed canes. It can then be potted on into a larger pot; the repotting may have to be carried out many times before the young plant is fit for planting in a permanent site out of doors.

In common with all young plants, the bamboo seedling requires not only warmth and good soil to grow in; it also needs adequate moisture. The watering of the young seedling should be carried out with great care, especially through the first few critical months of its life. The soft shoot enjoys a moist humid atmosphere; round about 70°F is ideal, and it will grow only slowly in lower temperatures. Do not overwater, and if possible stand the pots in a dish of slightly tepid water, and allow the moisture to soak through the soil, then drain the excess water out and stand the pot back in a warm corner. Never apply moisture in overhead soakings, as the first three months of growth are the dangerous ones, and it is during that period that the damping-off scourge can strike. Cheshunt compound can be used to forestall this happening, but if care is given in the application of water it should not be necessary.

⊀ 5 ⊁

Commercial Uses of the Bamboo

The edible bamboo shoot—Harvesting and curing canes—Paper pulp—Future possibilities

---

THE EDIBLE BAMBOO SHOOT

The young succulent shoots of the bamboo have for centuries been held in high esteem by all Asian peoples as an edible delicacy. Large quantities of prepared shoots are exported every year by China and Japan. Cultivation of the really large-diameter shoots suitable for market is impossible in most European countries, although the very warm and humid districts of southern France, Italy, and Spain might possibly be suitable for the purpose. There have been many reports of marketable quantities being grown in some of the south and south-eastern states of the United States of America, and judging by the increasingly large amount imported by that country every year a grower positioned in a favourable growing area should find a ready market for the home-grown shoot.

The English grower cannot be sufficiently sure of a warm spring, or of the very high early summer temperatures that are required to raise the shoots of the large-diameter species. This does not mean that the smaller-diameter shoots cannot be eaten; on the contrary, most, if not all, hardy species have shoots which are edible. In many cases the 'running' hardy types are superior in flavour to the large tropical species. The Phyllostachys give the thickest shoots amongst the hardy species, but do not supply the quantity, whereas the Arundinaria species supply the quantity but not the diameter. Most of the Phyllostachys shoots have a slightly bitter flavour which must be eliminated in the kitchen. *P. aurea* is an exception, for the shoots from this species are naturally sweet and free from any bitterness or acridity. Nearly all the Arundinaria shoots are semi-sweet.

I would like to emphasize that the edible shoots referred to are not the 'shoots' that the pandas are reported to eat. Panda fodder consists of the bamboo branches and foliage. I was obliged to point out this error to one television commentator who in his broadcast informed the viewers that the 'shoots' for the panda Chi-Chi were gathered in a remote district of Cornwall, and that the plant was so rare that he could not even name the district in case the plant was damaged. He insisted that the bamboo only grew in this remote part of the country, and nowhere else, but the bamboos shown on the screen were in reality two separate species, not one, and they were the common *Sasa palmata* and *Arundinaria japonica*. Both are well-known species which grow freely all over Great Britain, and could be found in the vicinity of the Zoo itself. It may not be common knowledge, but both species have become more or less naturalized in this country.

Many people ask whether the bamboo can be grown in the same manner as asparagus or mushrooms. This is not possible. Not only are the bamboos evergreens, but they depend on the new shoots to perpetuate the group. The only way to obtain edible shoots is to select a few from an established bed. The majority of the shoots must be left to grow into canes, and furthermore, if edible shoots are to be taken regularly, the bed will need a rich top dressing each year.

The edible-shoot harvest is taken during the spring or early summer, according to the species. As soon as the tips of the young spears begin to appear through the soil, select those which look the thickest and cover them over with a small mound of soft soil. This helps to keep the flesh white, and enables it to increase in bulk. If direct light does reach the shoot in its early stage of growth, the flavour will be marred, and will be bitter instead of sweet. Do not cut shoots which have begun to elongate above the soil surface. These are useless for culinary purposes. The flesh at that stage of growth is not only bitter, it is also tough and woody. When the tips of the young shoots show through the mound of soil the time is right for cutting. Pull the soil carefully away and dig down to the base where the shoot joins the parent rhizome. The cut must be made on the short link which connects the rhizome to the shoot. The rhizome itself should not be severed, as there may still be forward buds swelling along the line. Take care also where you tread, for many more shoots may be just under the soil surface, and could easily be damaged by clumsy feet.

Preparation of the succulent shoots for the table is simple and straightforward. Cut off the tough woody portion at the base, and slit through the sheath lengthwise with a sharp knife. Strip off the sheath until the white 'meat' is exposed. Cut off also any bits which are damaged or discoloured. The shoots can be cooked in various ways.

They can either be sliced thinly across the grain or diced into small cubes. These are boiled for twenty minutes in salty water. Then they are strained and simmered in a frying-pan, or fried in fat until golden brown. If the shoots are from a known acrid species, or have been exposed to daylight for a time, boil them in water for ten minutes, then in a change of water for another ten. Add salt to the last boiling, and this should remove any trace of bitterness. The cooked shoots can be used in combination with other vegetables, or alone with a meat dish, and are delicious if eaten with melted butter. The Chinese pickle them, but the method they use is unknown to me. If the shoots are from a sweet species, they can be sliced and used as an addition to a salad mixture. All bamboo shoots are firm and crisp in texture when eaten raw, and never become mushy when cooked. The flavour is subtle and singularly characteristic of this vegetable, but it is not strong. For this reason savoury ingredients are often incorporated in the recipe.

HARVESTING AND CURING CANES

Garden Canes

There is an almost unlimited demand in European and American countries for strong bamboo canes of various thicknesses and grades to be used as pea sticks, bean poles, plant supports, etc. The vast majority of the canes now imported come from the Far East, where they are more often than not obtained from the Tonkin bamboo (*A. amabilis*). They are expensive, and the thrifty gardener looks after them carefully, but it is still not widely realized that many of the hardy bamboos grown in Europe and in America produce canes of good quality. These canes are quite durable and have a market value. The cultivation of bamboo canes for commercial purposes is profitable in the U.K. and in most European countries. The intending grower must, however, choose the species to suit his needs, and he must also watch two points before selling his crop. Firstly, the canes should only be cut when they are ripe. Secondly, the cut canes *must* be cured thoroughly. Old dead canes are brittle and snap easily, and under-ripe canes are too soft, but home-grown bamboo canes, when harvested in the right condition and cured throughly, are equal to the imported material and, more to the point, they must be cheaper. A bamboo cane takes three years to reach maturity, and canes under that age have not had time to lignify. To facilitate identification of the different yearly crops, it is a good idea, when dealing with the more valuable varieties, to mark each cane of the

Bamboo Shoots (*Arundinaria*)

current crop with a dab of paint. Each succeeding year's crop can be marked in a different colour. Any oil-based or plastic paint will last long enough in the open air until the canes are fit for harvesting, and the residue of the paint can easily be removed during the curing process.

The ripe canes should be severed as near the ground as possible, and just below the lowest joint or node. Try to avoid cutting the thinner types between the joints, as this tends to split the cane. A pair of strong secateurs will cut canes of medium thickness, but those of larger diameter may have to be cut with a pair of heavy-duty branch cutters. Curved or distorted canes should be discarded unless required for a special purpose. These canes can be straightened, but the various processes, such as soaking in water, or the direct application of heat, are time absorbing, and unless the cane is of outstanding value or rarity they are not worth while. All branches must be cut off close to the stem, using a fine-bladed hacksaw on the large ones. Then discard the fine whippy top. The curing process is lengthy and must not be hurried. If there are ample storage facilities, the cut canes should be laid flat on a rack equipped with supports to keep the canes from bending, or laid on a flat, preferably wooden floor in a well-ventilated shed. The canes must be turned over occasionally to assist the curing process. Adequate ventilation is necessary and damp conditions should be avoided. Canes of medium thickness, up to 1 inch in diameter, should be left to dry for at least six months, and twice as long is needed by the really thick types. When cured, they can be bundled into portable quantities and stacked base to tip in an upright position. The curing process is slow, but essential, as it dries out not only the hard outer surface of the cane but also the soft inner section. It will be found that the mature cured cane is lighter than the freshly cut one. This is not a sign of weakness but of strength. The gradual tightening of the cane fibres gives the natural suppleness and pliability, plus strength, for which the bamboo is famous. Many schoolboys can verify that statement.

The most ambitious attempt to replace imported canes by the home-grown kind is that started at Lanivet in Cornwall by British Bamboo Groves Ltd., during World War II. The variety selected for the most extensive planting was *A. japonica*, and today the groves of that species from which the harvest is taken at Lanivet cover many acres. Other species tried there, but with less success, include *A. anceps*, *A. fastuosa*, and *S. palmata*. The yearly output is reported to be in the region of two and a half million canes. The demand for the home-grown product is insatiable, and still exceeds the supply. The United States Government is now sponsoring bamboo groves for commercial purposes, but the British Government has so far failed to do so.

I myself have never grown canes commercially, and cannot therefore pronounce on the yield per acre, or on the labour costs, marketing, or profits, etc., but having grown many species outdoors, and cut many hundreds of canes for sundry garden purposes, I can say with conviction that a high-quality cane can be produced quite easily by any grower in the United Kingdom. The best garden canes come from species of Arundinaria. Many of these species throw tall, rigid, and polished canes in abundance every year. A clump of *A. anceps*, for instance, may have half a dozen canes after the second season from planting, and could have as many as thirty or even forty by the end of the fourth year. With periodical division and subdivision of the offsets the original clump can easily become a dense grove of tall canes in a few years. Every garden in the temperate zone could grow its own canes, have a surplus to sell, and still maintain generous cane brakes for screening or decorative uses.

My personal experience is that *A. japonica*, *A. fastuosa*, and *A. anceps* give good straight garden-quality canes, so also do *A. niitakayamensis* (Pitt White clone), *A. simonii*, and *A. tessellata*. The slender but strong canes of *A. nitida* are also useful when a finer and less obtrusive plant support is required. Canes from this species can also be woven into hurdles and plant-protection screens.

PAPER PULP

In the Orient, for centuries in the past, the bamboo, and especially *P. pubescens*, furnished the basic material for the manufacture of paper. Today *S. palmata* is also used in great quantities for processing into modern hardboard both in China and Japan. In recent years new factories have been built in Formosa and India for this industry. In India and Pakistan the usual source is *Dendrocalamus arundinaceus*, a rapidly growing tropical species which can be cropped every three or four years, yielding, it is reported, three-quarters of a ton of cut cane annually per acre.

Chemically the bamboo contains four principal substances, starch, pectin, lignin, and cellulose. The last-named substance is the basic material used in the manufacture of paper, and as the reader may be interested to learn how cellulose can be obtained from the raw bamboo, the following is a brief outline of the process employed. The woody stems are crushed between heavy iron rollers to break down the tough joints and separate the fibrous bundles so that the chemical solutions next to be introduced can penetrate the mass satisfactorily. The crushed material is then cooked in solutions, and at temperatures and pressures, which vary from one process to another. For example, the mass can be boiled with weak caustic soda for several hours under a pressure of six or seven atmospheres. Sulphites, sulphides, and sulphates are also employed, the object being the same—to dissolve out the lignin, pectin, starch, and unwanted minority constituents from the principal material, cellulose, which is resistant to these chemicals, and is to constitute the residual pulp for the manufacture of paper.

The suitability of bamboo pulp for paper-making is largely dependent on its fundamentally fibrous structure. Measurements show that the fibres in the original macerated cane are between one-twelfth to one-seventh of an inch in length, but the pattern of the invisible structure of the cellulose component is probably of greater importance still. Investigation of the intrinsic structure indicates that the cellulose molecule consists of a chain of about a hundred glucose-like units, joined end to end in a filamentous fashion common to all those compounds which in recent years have proved suitable for the manufacture of man-made textile fibres. The quality and the individual characteristics of the paper finally produced are, of course, greatly affected by the species of bamboo selected, and by the details of the cooking process employed.

As the population of the world increases yearly, and the standards of education in every country reach new heights, so does the corresponding demand for paper pulp. The massive drain on soft wood, the main source of paper pulp, continues, and in the near future the demand may well exceed the supply. Sources of alternative materials must then be found, and the bamboo could supply a high-quality, yet cheap substitute. With this purpose in mind, trials were attempted in the United Kingdom during the last world war, when imported wood pulp was in short supply, but were abandoned at the close of hostilities. It may be that at a distant date, when the botanists and horticulturists have done more research into hybridization, we may see bamboos grown in this country especially for paper production, with names such as 'Bowaters seedling' or 'Spicers select'.

FUTURE POSSIBILITIES

Some species of hardy bamboo are very rampant in growth. This often unwelcome feature is put to good use in some parts of the world, particularly in the High Himalaya countries (Sikkim, Nepal, Bhutan, etc.), where several such species originated. In those districts some species, such as *A. racemosa*, appear as rather oversized grasses, and the growth is cut and harvested annually for cattle fodder, in much the same way as the farmer gathers in meadow grass in other lands. In temperate countries few species apart from *A. tecta* have been used in this way. This North American species, commonly known as the 'switchcane' or 'smallcane' has been the subject of trials in Washington County, North Carolina, and has been shown to rival, when grown under favourable conditions, the highest-quality American grasses. *A. tecta* and probably such species as *S. pygmaea* could well be given a trial in other countries for the same purpose.

A future in the building industry?

Since the advent of plastics, fibreglass, and other synthetic materials, the popularity of the bamboo has suffered a decline in many manufacturing fields. It is still used to make high-class walking-sticks, and the rhizomes of *P. nigra* still supply the handles for quality umbrellas. The trout fisherman continues to hold the split cane rod in high esteem for its outstanding lightness, elasticity, and strength. *P. aurea*, the 'fishpole' bamboo, is recommended in the United States of America in the manufacture of walking-sticks and umbrella handles. However, a new vista may soon open in the building industry for concrete reinforcement. In the United States of America, the Engineering Experimental Station at Clemson, North Carolina, reports that trials show that concrete beams with longitudinal bamboo reinforcement can be designed which can safely carry loads from two to three times as great as those expected from unreinforced concrete members of the same dimension. It is stressed that large-diameter, mature, and fully seasoned canes must be used. No firm conclusions have yet been published, but it could possibly be that the bamboo has once more surprised us by making itself useful in an entirely new dimension.

Descriptive Classification of the Species, Varieties, and Cultivars

*Arundinaria—Phyllostachys—Sasa—Bambusa—Chusquea—
Shibataea—Lesser-known hardy groups*

The woody-stemmed perennial grasses collectively known as bamboos belong to the family Poaceae (Gramineae), and form the sub-family Bambusoideae. In this sub-family there are several hundred species, of which roughly ninety have so far been proved to be hardy and suitable for open garden cultivation in temperate climates. All the hardy species are evergreen (except *A. tecta* var. *decidua*), renewing their foliage each spring, and bearing aerial stems or canes which vary in height from pygmies of less than 12 inches in height, to giants of 50 feet or more.

Most of the species which are readily adaptable to variable weather conditions in temperate climates are members of the genera of Arundinaria, Phyllostachys, Sasa, plus a few smaller and lesser-known groups. The Arundinaria has been divided into several distinct genera by modern botanists, notably the Japanese. Some of the divisions are:

Arundinaria proper
Chimonobambusa
Indocalamus
Nipponocalamus
Pleioblastus
Sasamorpha
Semiarundinaria
Sinarundinaria
Thamnocalamus, etc.

There is, however, still some difference of opinion amongst those who have initiated these new genera, and several of the divisions, notably Nipponocalamus, Sasamorpha, and Pleioblastus, have yet to be recognized in many countries. Because of the divergence of views held by botanists concerning the exact status of the genera in question, I have therefore classified the various species, cultivars, etc., in this chapter under the names by which they are generally recognized, but to avoid controversy I have included under each heading a list of the other known names or synonyms. Included also is the name of the country of origin and native names if known, plus any known facts of the history of the species concerned, together with such relevant data which may be of interest to the reader. In order not to be involved in the battlefield of formal bamboo taxonomy, I have omitted the partial and double citation required by the International Code of Botanical Nomenclature, as I believe that they are necessary only to taxonomists, and tend to complicate further the already complicated bamboo nomenclature.

The first section of the chapter contains those hardy bamboos which are under general cultivation in the United Kingdom, the second section containing those bamboos which, though proved as hardy, are less well known. Each section is headed by the genus Arundinaria, followed by Phyllostachys, Sasa, Bambusa, and ending with the lesser miscellaneous groups. All are listed in alphabetical order. The cane heights and diameters recorded are what I believe to be a fair mean average measurement of hardy bamboo canes when grown to maturity in temperate zones. As the conditions in the southern part of England are fairly average, I have decided to use the gardens at Pitt White as the norm. The heights and diameters given may in some cases conflict with those given in other bamboo literature, because growth is greatly influenced by local conditions of soil and climate, as has been explained in a previous chapter. I do not intend to imply that the heights, etc., which I have given are the maximum growth of each bamboo species—on the contrary, nearly every species can grow canes of much greater size in its native habitat than those produced in temperate countries. There are exceptions; for instance, many of the bamboos which come from the high Himalayas, which grow much taller canes in the temperate zones than they do in their native homelands.

As the cane height and diameter vary in different localities, so, frequently does the size of the leaves. The leaf measurements of some species show a considerable range of variation on canes growing in the same clump, and the measurements given are intended as a general rule or guide and a reasonable average. They indicate relative sizes as between one species and another. Apart from these variations other botanical features remain constant. Some of the features, such as the number of secondary veins in the leaves, the leaf colour, the number of branches borne per node, colour of the fresh shoots, etc., are of immense value in helping one to distinguish one species from another.

LIST OF ESTABLISHED HARDY BAMBOOS

Arundinaria anceps
       angustifolia
       auricoma
       chino
       disticha
       falcata
       falconeri
       fastuosa
       fortunei
       gauntlettii
       graminea
       hindsii
       hookeriana
       humilis
       japonica
       laydekeri
       macrosperma
       marmorea
       murielae
       niitakayamensis (Pitt White clone)
       nitida
       pumila
       pygmaea (glabrous form)
       pygmaea (pubescent form)
       racemosa
       simonii
       spathiflora
       tecta
       tessellata
       vagans

Phyllostachys aurea
       bambusoides
       boryana

Arundinaria

    castillonis
    flexuosa
    fulva
    henonis
    marliacea
    mitis
    nigra
    puberula
    pubescens
    punctata
    quilioi
    sulphurea
    violescens
    viridi glaucescens

Sasa chrysantha
    palmata
    palmata var. nebulosa
    senanensis
    tessellata
    veitchii
    veitchii var. nana

Bambusa quadrangularis
Chusquea culeou
Shibataea kumasasa

Arundinaria

The obvious recognizable differences between the Arundinaria and the other types of hardy bamboo, excluding the finer morphological characters of inflorescence, etc., are:

the round, smooth, upright, reedlike canes;
the lack of obvious internodal grooving;
the sequence of branch emergence, which is from top to bottom;
the cane sheaths, which adhere to the middle of the base after they dry, and in some cases may remain attached to the cane for many seasons;
the rootstock action, which is invariably active and often rambling.

73

Arundinaria anceps
Country of origin: India

A. anceps is an extremely hardy bamboo, but in some very cold districts it may lose foliage during a hard winter. The canes and the rhizomes seldom come to harm, and in the following spring there will always be abundant foliage to replace the fallen leaves. The canes of this species, if ripened properly, are ideal for garden use. This species is also recommended as a good hedging and screening type, and although the rhizomes are inclined to wander, the straying shoots can be broken off easily when young and brittle. Named as *anceps* by A. B. Freeman-Mitford, who dubbed it the 'doubtful' bamboo, as at the time of its introduction into England there was some doubt as to its true country of origin. It has now been definitely established as an Indian bamboo from the north-west Himalaya mountain range, where it grows at altitudes of between 10,000 and 11,000 feet above sea-level in the states of Sikkim, Bhutan, and Gharwal. It was introduced originally into the United Kingdom by a Colonel E. Smyth in 1865, and first cultivated in the gardens at Elkington Hall in Lincolnshire.

A. anceps flowered in England in 1911, and again in the 1935/6 seasons, and was reported as having flowered at Dereen in Eire in 1955/6. Information received stated that the clumps did not die out, and at the time of writing are once more back to full stature.

DISTINGUISHING FEATURES The well-spaced, straight, smooth, and glossy deep green canes, especially noticeable when they arch at the summit with the heavy weight of the fine foliage to form imposing plumes. The only bamboo which bears a superficial resemblance to *A. anceps* in its early stages of growth is *A. niitakayamensis*, but the latter eventually grows much taller in all districts.

CANES 12 feet plus in height, by up to $\frac{3}{4}$ inch in diameter. Erect in habit at first, but may cascade in the succeeding seasons if grown in a moist and wind-free environment. Bloomed fairly heavily with a fine bluish-white powder above and below the joints, but this quickly disperses. Deep glossy green at first, maturing to a dull greeny brown shade.

NEW SHOOTS Appear from late May onwards. Smooth. Tipped with deep olive green and edged with pale purple.

CANE SHEATHS Not persistent, except on the later-emerging canes, and these may remain attached until the following spring. From 4 to

8 inches in length. Highly glossed on the interior surface. Beige green at first, quickly fading to a dull straw colour.

NODES Not very prominent. Ring compact, thick, and straw coloured. Variable distances between nodes, but can be as far apart as 14 inches.

BRANCHES The larger canes do not carry branches on their lower 4 feet. Those at the tip are the first to appear, the lower ones opening in succession. Slender and arching. Three to four branches per node in the first season, multiple with numerous twigs thereafter.

LEAVES Average 4 inches in length by $\frac{1}{2}$ inch wide. Occasionally leaves may appear which are up to 6 inches in length, but these only appear on the younger canes. Midrib fine. The base is bluntly wedge-shaped, and the tip short and finely tapered. Leaf-stalk short. One margin fringed with fine bristles. Tessellation good. Four to six pairs of secondary veins. Glossy mid-green on the upper surface, dull greyish green on the underside.

PROPAGATION Basal cane cuttings. Division of the clump.

ROOTSTOCK Running, an uncommon type—essentially caespitose, but with long running axes branching off from the short rhizomes of the standing culms. They are segmented and sheathed, but are devoid of any kind of lateral buds, and each ends in a short typical rhizome from the tip of which a new cane springs ready to multiply itself by tillering. These axes act as barren and rootless connecting links from the side of one rhizome to the base of another—a kind of grossly elongated rhizome neck. The resultant growth above ground is a compact clump surrounded at a short distance by scattered groups of new canes.

Arundinaria angustifolia
Syn. *Bambusa angustifolia*
    Bambusa vilmorinii
Country of origin: Japan

A. angustifolia is a hardy, semi-erect bamboo, and has the distinction of being the only bamboo which possesses leaves which are of the same colour on both surfaces. A rampant grower, it is inclined to get out of hand quickly; therefore should be planted in a position where it can be allowed plenty of room to expand. A native of Japan, it was introduced into England by way of France *circa* 1895. Given the name *angustifolia* or 'narrow leaved' by A. B. Freeman-Mitford.

DISTINGUISHING FEATURES The smooth narrow leaves, and the unique similarity of the colouring on both leaf surfaces.

CANES 6 to 8 feet in height by ¼ inch in diameter. Thin walled. Light green in colour, deepening to purple when mature.

NEW SHOOTS Appear from late May onwards. Bright green tipped purple.

CANE SHEATHS Not very persistent. Slightly downy, with an erect tuft of hairs at the tip. Dull buff coloured.

NODES Rather prominent. Average 10 inches apart, but crowded closer together towards the tip of the cane.

BRANCHES Usually two to four per node. Erect and slender.

LEAVES Up to 6 inches in length by ¼ inch wide. Smooth in texture. Rounded at the base, and tapering to a sharp point at the tip. Edged on both margins with fine bristles. Leaf-stalk short and stout. Two to four pairs of secondary veins. Brilliant green on both leaf surfaces. Often known to sport leaves with white variegations.

PROPAGATION Rhizome cuttings. Division of the clump.

ROOTSTOCK Running, inclined to become invasive, soon forms dense and thicket-like clumps.

Arundinaria auricoma
Syn. *Arundinaria viridistriata*
    Bambusa fortunei var. *aurea*
    Pleioblastus viridistriatus
Country of origin: Japan

A. auricoma is a very hardy bamboo, and is by far the best of the variegated species. It is the ideal bamboo for growing in the ornamental border, or in a sunny position in a rockery. It also makes a very good tub plant. In fact, it is a good plant for any position in the garden where it can be allowed to display its beautiful foliage to its full advantage. The colouring is even more striking when the leaves are wholly yellow, and this often occurs. If the canes are cut back to the ground in the autumn the leaf colouring on the new canes is fresh and sparkling. Named as *auricoma* or 'golden haired' by A. B. Freeman-Mitford. Believed to have been introduced into Europe in the 1870s. It has flowered frequently over the last fifty years in France, Belgium, and in the United Kingdom. The flowering is usually partial, a few canes bearing flowers, the rest normal leaves.

DISTINGUISHING FEATURES The brilliant golden-yellow variegated foliage is outstanding, making this species instantly recognizable from the other yellow variegated species by its sheer brilliance.

CANES Up to 6 feet in height by $\frac{3}{4}$ inch in diameter when grown under ideal open conditions, but much smaller when grown in a shady position. Thick walled for the size of the cane. Dark purplish green.

NEW SHOOTS Appear from late March onwards. Pale creamy gold, with purplish shading at the tips.

CANE SHEATHS Persistent. Hairy at the base, and fringed with minute hairs at the apex.

NODES Not prominent. From 3 to 6 inches apart. Ring dark and lustrous.

BRANCHES Usually borne singly, but occasionally in pairs. Long in relation to the height of the cane.

LEAVES Very variable in size, and can often be seen as large as 8 inches in length by $1\frac{3}{4}$ inches wide. Rounded at the base, and pinched in sharply about half an inch from the tip to end in a short sharp point. Leaf-stalk short and squat. Midrib fairly prominent and palpable for half of its length from the base. Edged with fine bristles on both margins, but often only partially so on one side. The underside of the leaf is smooth to the touch, this being caused by the close felting of fine white hairs. The upper surface is rough to the touch, progressively growing rougher as the leaf ages. Tessellation rather prominent. Five to seven pairs of secondary veins. Pea green, striped with golden yellow on the upper surface, similar but duller on the lower. The golden striping varies in both width and number from leaf to leaf.

PROPAGATION Division of the clump.

ROOTSTOCK Running, not invasive, and can easily be kept under control.

Arundinaria chino
Syn. *Arundinaria maximowiczii*
     Arundinaria simonii var. *chino*
     Bambusa chino
     Nipponocalamus chino
     Pleioblastus chino
     Pleioblastus maximowiczii
Country of origin: Japan
Native names: Azuma-nezasa; Shinegawadake

A. chino is a very hardy, but not a good-looking bamboo, as the foliage tends to be rather dull and drab, especially during the winter months. In good soil conditions it can spread rather rapidly, and should be confined to the poorer soil areas or in an isolated position. The rapidly growing semi-dwarf canes can, however, be used to some advantage in screening slopes and banks.

Found originally in the Honshu district of Japan, it has been well established in China for many years. Introduced into English gardens in 1875.

This species has flowered freely in Japan and China, in many parts of Europe, and in several areas bordering the Gulf of Mexico in the United States of America. The Curator at the State Nikita Botanical Gardens at Yalta (A. M. Kormilitzin) states that the species flowered there in 1962/3, and although the flowering was profuse, no seed were set. The clumps did not die out, and fresh canes grew freely in the following spring.

Although believed by some authorities to be a garden variant of *A. simonii*, the seed from this species, when germinated, remain true to type.

DISTINGUISHING FEATURES The dark green, purplish-blotched canes. The leaves, which are borne in stiff plumes, not unlike a fox's brush.

CANES Up to 7 feet in height by up to $\frac{1}{2}$ inch in diameter. Dark green, usually blotched and stained with purplish markings.

NEW SHOOTS Appear from late May onwards. Olive green, lightly edged with pale purple.

CANE SHEATHS Fairly persistent. The inner surface is highly glazed. Pale purple when young, maturing to a dull straw colour.

NODES Prominent. Usually from 8 to 10 inches apart.

BRANCHES Borne singly, but occasionally up to five on the lower part of the cane, multiple towards the tip. Long in proportion to the height of the cane.

LEAVES Up to 10 inches in length by $\frac{3}{4}$ to 1 inch wide. Bluntly rounded at the base, and tapering to a long slender point which ends in a fine sharp tip. Leaf-stalk short and pale yellowish green in colour. Midrib fine, but palpable for half of its length from the base. Tessellation good. Four to six pairs of secondary veins. Dark green on the upper surface, dull greyish green on the lower.

PROPAGATION Division of the clump. Rhizome cuttings.

ROOTSTOCK Running, but not invasive, except in good growing conditions.

Arundinaria disticha

Syn. *Arundinaria argenteostriata* var. *disticha*
    Bambusa disticha
    Bambusa nana
    Nipponocalamus argenteostriatus cv. 'Disticha'
    Pleioblastus distichus
    Sasa disticha
Country of origin: Japan
Native name: Oroshima-chiku

Given the name *disticha* (arranged in two rows) by A. B. Freeman-Mitford, it is also known as the 'fernleaf' or 'dwarf fernleaf' bamboo in the United States of America, where it is grown extensively in many areas. The paired leaves do, in fact, resemble fern leaves, but when this species is cultivated in a shady position the branches tend to elongate and the singular leaf formation becomes more open. *A. disticha* is an elegant species, and an attractive addition to the dwarf garden bamboo collection, where it is particularly suitable for growing in rockeries. Introduced into Europe from Japan in the late 1860s, it has been in cultivation in United Kingdom gardens since 1870.

DISTINGUISHING FEATURES The small leaves, and especially the curious arrangement of them, gives this pretty little bamboo a peculiar and distinctive appearance.

CANES 1 to $2\frac{1}{2}$ feet in height by $\frac{3}{4}$ inch in diameter. Bright green, often shaded purple.

NEW SHOOTS Appear from late May onwards. Olive green tipped purple.

CANE SHEATHS Not persistent. Downy at first. Dull dun coloured.

NODES Not prominent. Vary in distance from 4 to 6 inches apart.

BRANCHES Normally borne singly at each node, but occasionally can be found in pairs.

LEAVES Up to $2\frac{1}{2}$ inches in length by $\frac{3}{4}$ in wide. Arranged in two opposite rows. Leaf-stalk short. Slightly downy on both surfaces, especially towards the base. Edged with fine bristles on both margins. Tessellation good. Three to four pairs of secondary veins. Bright green on the upper surface, dull matt green on the lower.

PROPAGATION Division of the clump. Rhizome cuttings.

ROOTSTOCK Running. A fairly active mover, can become moderately invasive under good growing conditions.

Arundinaria falcata
Syn. *Bambusa falcata*
    Bambusa gracilis
Country of origin: India

An attractive but rather delicate bamboo, *A. falcata* is only suitable for growing outdoors in the milder areas, or in a conservatory in colder districts. Originally comes from the mountain states of Sikkim and Bhutan, where it grows at altitudes of between 6,000 and 8,000 feet above sea-level.

According to E. G. Camus, in his book *Les Bambusees*, this species flowered in France and in England, but the dates of these flowerings are not known. Later records show that it flowered at Kew, and at the National Botanic Gardens at Glasnevin in Eire in 1926, and again at Kew in the 1938 and 1951 seasons. Munro describes this species as belonging to a group of four, which includes *falcata, hookeriana, intermedia,* and *khasiana.* In the mountain regions of the Himalayas they bear distinctly different leaf- and flower-bearing canes annually. These die down under the winter snows and fresh growth appears in the following spring. When these species are grown in milder countries the annual dying of the canes does not occur, thus each succeeding crop of canes tends to grow taller than its predecessor, until a good stand of mature canes evolves. But this species must always be treated strictly as a 'mild area' bamboo, and should always be given a warm and wind-free environment.

Introduced into the United Kingdom by Nees, *circa* 1870, and named by him as *falcata* (sickle-shaped).

DISTINGUISHING FEATURES The absence of tessellation in the leaves, the deep crimson coloured new shoots, and the greyish-green colour of the thin-walled canes, which grow closely erect on the inside of the clump, whilst the ones on the outer edge branch outwards gracefully.

CANES 20 feet plus in height by $\frac{1}{2}$ inch in diameter. Greyish green, ripening to a yellowish-brown shade.

NEW SHOOTS Appear from late May onwards. Deep crimson, fading to a dull purple shade. Edible.

8. *A. falconeri* at Mount Usher, Co. Wicklow, Eire. By courtesy of the
Irish Tourist Board

9. *A. auricoma*, Pitt White, Uplyme, Devon

CANE SHEATHS Not persistent. Margins edged with fine bristles. Long, often covering the internode. Striated and coarse in texture. Highly polished on the inner surface. Purple crimson when young, but fade quickly to a dull straw shade.

NODES Fairly prominent. Average 1 foot apart on mature canes. Coated with a velvety down.

BRANCHES Numerous at every node. The thin new branches sometimes force their way through the base of the sheath, often tearing it into strips.

LEAVES Up to 6 inches in length by $\frac{1}{2}$ to 1 inch wide; the terminal leaves are often much larger. Bluntly tapered at the base, short pointed at the tip. Thin and papery in texture. Fringed on both margins with fine bristles, but on one side more than the other. Leaf-stalk medium length, and furrowed on the upper portion. No visible tessellation. Two to six pairs of secondary veins. Pale to mid-green on the upper surface, dull greyish green on the lower.

PROPAGATION Division of the clump. Basal cane cuttings.

ROOTSTOCK Caespitose. Does not sucker, but grows from the main clump slowly. Makes an exceptionally fine tub plant.

Arundinaria falconeri
Syn: *Arundinaria nobilis*
    Bambusa floribunda
    Thamnocalamus falconeri
Country of origin: India

Although not one of the hardiest of bamboos, *A. falconeri* can be grown fairly easily in warmer and more sheltered areas. It prefers to grow in semi-shade, as do all the thin-leaved species. A native of the north-east Himalayas, this species was introduced into the United Kingdom by a Colonel Madden in 1847.

The thin-walled canes are very pliable, and are used extensively in the manufacture of fishing-rods, and in many parts of India for weaving baskets.

There has been some argument by various authorities as to whether the synonym *nobilis* given to it by Freeman-Mitford refers to the same species, and at least one grower in Cornwall insists that *falconeri* and *nobilis* are two distinctly different species. Further investigation by competent botanists is necessary to determine the truth of the assertion.

Clumps of this species have been recorded as having flowered in many areas of Europe, Algeria, and Sikkim in 1876, in Darjeeling in 1890, and in England in the 1929 and 1935/6 seasons. It sets seed freely and the germination percentage is reported to be very high. A very large clump of this species in the gardens at Abbotsbury in Dorset grew from seed produced from a previous clump. It is on record that the clump grew naturally in the soil at the base of the parent plant. At the moment of writing this same clump is in full flower, and a magnificent sight it is, too.

DISTINGUISHING FEATURES The absence of visible tessellation in the leaves. The brownish-purple stain at the joints, especially prominent on the older yellowish canes. The thin cane sheaths which are truncated at the summit. The very small, delicate, and papery thin leaves.

This bamboo grows into a compact, sharply defined shape made up of strong but relatively slender canes which are densely packed and rise to a height of perhaps 30 feet in good conditions. The outer canes curve gently outwards and the inner ones stand stiffly erect. They are generously clothed almost from the base with rich foliage carried on a multitude of fine branches and branchlets. It well deserves the name *nobilis* given it by Freeman-Mitford.

CANES 30 feet plus in height by up to $1\frac{1}{4}$ inches in diameter. Olive green at first, with a brownish-purple stain at the nodes. Matures to a dull yellow shade when grown in a sunny position.

NEW SHOOTS Appear from late May onwards. Pale crimson fading to a dull purple shade.

CANE SHEATHS Not persistent. Short. Paper thin. Truncated at the summit. Highly polished on the interior surface. Pale crimson fading to straw.

NODES Fairly prominent. Average 10 inches apart. Stained with a brownish-purple hue which extends roughly $\frac{1}{2}$ inch on either side of the ring. This colouring is intensified when the cane is exposed to full sunlight. Coated at first with a whitish bloom which disperses quickly. Ring thin and straw coloured.

BRANCHES Bears numerous branchlets at every node from the first season onwards. Slender and smooth textured. Five to seven leaves per branchlet.

LEAVES Up to 4 inches in length by $\frac{1}{2}$ inch wide. The terminal leaves may be much larger. Thin and papery in texture. Bluntly wedge-shaped at the base, and short tapered at the tip. Midrib fine but palpable. No discernible tessellation, but furnished with numerous translucent glands. Three to four pairs of secondary veins. Brilliant medium green on the upper surface, matt medium green on the lower.

PROPAGATION Division of the clump. Basal cane cuttings.

ROOTSTOCK Caespitose. Does not sucker, but grows from the main clump very slowly. Makes an excellent ornamental plant when grown in a shady spot in warmer gardens. It is especially recommended as a tub species for the conservatory.

Arundinaria fastuosa
Syn: *Arundinaria narihira*
  Bambusa fastuosa
  Semiarundinaria fastuosa
Country of origin: Japan
Native name: Narihiradake

One of the tallest growers of the Arundinaria, *A. fastuosa* was named by M. Latour Marliac of Temple sur Lot in France. It was introduced by him from Japan into France in 1892, and was brought to England in 1895.

The habit of growth is stiff and upright, making it stand out head and shoulders above most of the other hardy bamboo species. The name *fastuosa* (stately) is exactly the right name for this elegant bamboo. Hard weather never seems to bother it; in fact, the stands growing in the gardens at Pitt White stood up well to the severe winter of 1962/3, when many shrubs, including many of the hardy species, perished. In the United States of America, where this bamboo has been cultivated for many years, it is reputed to have survived temperatures as low as $-4°F$.

A native of the Honshu, Shikaku, and Kyushu districts of Japan, it is known under the name of Narihiradake, after Narihira, a hero in Japanese mythical romance of the eleventh century.

The only other hardy bamboo with which it is liable to be confused in its early stages of growth is *A. simonii*, but *A. fastuosa* is a much more elegant grower, neater and more compact in habit, and has the easily distinguishable claret colouring on the inside of its cane sheaths.

In recent years it is on record as having flowered in England in the 1935/6 seasons, and in Japan in 1951. Several clumps at Pitt White have been flowering sporadically for many years.

The flowering habit of this species is unique, so unusual, in fact, that it contradicts the established botanical identifying features which apply to the Arundinaria. The canes, particularly the later-flowering ones, show, from the branched section upwards, distinct Phyllostachys characteristics. i.e. deeply flattened or indented branch grooves, combined with the familiar Phyllostachys zigzag manner of growth. On the lower branchless section the cane is smoothly columnar, and perfectly cylindrical, with not even a trace of grooving. In fact, when in flower this species combines the visual characters of both genera on each single cane. This unique occurrence appears to be confined to *A. fastuosa*, and it certainly does not happen on the flowering canes of *A. hindsii* or *A. simonii*, both of which have flowered here recently.

DISTINGUISHING FEATURES Easily identified by the tall, very upright and stately canes. Also by the unusual claret colour on the interior surface of the mature cane sheaths. Distinguished from var. *yashadake* by the more dense arrangement of the branches, and by the length of the prophyllum, which is $\frac{1}{2}$ inch on *A. fastuosa*, whilst on var. *yashadake* it is from 1 to 2 inches long.

CANES 25 feet plus in height by up to $1\frac{3}{4}$ inches in diameter. Very thin walled for the size of the cane, and easily split. Ramrod straight. Deep glossy green, but purplish tinged when young and maturing to a yellowish-brown shade.

NEW SHOOTS Appear from late April onwards. Smooth and glossy. Tinged deep purple at the tip. Edible.

CANE SHEATHS Not persistent, except on the later-emerging canes, where they may remain attached until the following spring. Can be as long as 10 inches by 4 inches wide. Thick textured. Downy. Purplish in colour on the outside, claret tinged on the inner surface. Quickly fades to a dull straw shade.

NODES Not prominent. Often up to 18 inches apart.

BRANCHES Two, sometimes three on the lower part of the cane, but multiple on the higher limits. Growth stiff and erect. Short for the length of the cane. Normally four to six leaves per branch.

LEAVES Up to 10 inches in length by 1 inch wide on the younger canes, but much smaller on the mature ones. Leaf-stalk fairly long, pale yellow in colour. Constricted at the tip, and ends in a sharp slender point, wedge-shaped at the base. Lacks auricles. Tessellation clear. Edged with fine bristles on both margins. Six to seven pairs of secondary veins. Brilliant green on the upper surface, duller greyish green on the lower.

PROPAGATION Division of the clump. Basal cane cuttings. Rhizome cuttings.

ROOTSTOCK Though botanically classed as running, it only wanders at the edges. Makes a magnificent solo specimen plant.

Arundinaria fortunei

Syn: *Bambusa fortunei variegata*
     Pleioblastus variegatus
Country of origin: Japan

A. fortunei is the best of the white variegated bamboos, but though perfectly hardy it may lose some of its white-streaked foliage in a hard winter. The canes and rhizomes are seldom damaged, and fresh canes and foliage will break out anew in the following spring.

It was once believed that this bamboo was the first of the hardy species to be introduced into Europe, but it is now known that *P. nigra* has that honour, having been introduced in 1827, whilst it was not until 1863 that *A. fortunei* was introduced into Belgium by M. Van Houtte of Ghent.

This species is particularly suitable for rockery work, and also makes an attractive tub or pot plant.

DISTINGUISHING FEATURES The dwarf habit of growth. The bright clean white variegated markings of the leaves.

CANES Up to $4\frac{1}{2}$ feet in height by $\frac{1}{4}$ inch in diameter. Much smaller when grown under shady conditions. Pale green.

NEW SHOOTS Appear from late March onwards. Thin and long pointed at the tip. White with green tips.

CANE SHEATHS Persistent. Thick textured. Pale straw coloured.

NODES Not prominent. Usually 6 inches apart when mature.

BRANCHES Borne singly as a rule at each node, but occasionally seen in pairs. Slender and long for the size of the cane.

LEAVES Up to 8 inches in length by 1 inch wide. Bluntly rounded at the base, and pinched in to end in a short sharp point at the tip. Covered with fine white hairs on both surfaces, and particularly so on the underside. Midrib prominent. Leaf-stalk short, wide, and yellow or whitish yellow in colour. One margin edged with fine bristles, the other only partially so. Tessellation very prominent. Four to five pairs of secondary veins. The upper surface of the leaf is dark green striped with white, fading to paler green. The lower surface is duller green and shows the white striping less. Occasionally leaves can be found which are wholly white.

PROPAGATION Division of the clump. Rhizome cuttings.

ROOTSTOCK Running. Can be moderately invasive under warm conditions.

Arundinaria gauntlettii
Syn: *Sasa gauntlettii*

A. gauntlettii is a hardy dwarf bamboo, the original habitat of which is still obscure, but it is believed to be a native of Japan. Very little is known about this species, but an old catalogue of Messrs. Gauntlett Ltd. (Chiddingfold), Surrey, describes this bamboo as being a very distinctive dwarf species, with canes from $1\frac{1}{2}$ to $2\frac{1}{2}$ feet tall, purple in colour, with leaves which resemble a miniature *A. japonica* (?). A later prewar catalogue from the same firm states that 'it reaches 5 to 6 feet in height, and forms tall and spreading masses'.

Hillier and Sons, the nurserymen, of Winchester, Hants, have informed me that the clump planted by them in 1959 has only grown to $2\frac{1}{2}$ feet in height and only covers an area $2\frac{1}{2}$ feet square since planting. The clump at Pitt White has achieved practically the same height and spread as Hilliers'. However, it is believed to be fast growing, and the closely packed canes make good ground cover, but more botanical investigation is required before these facts can be proved.

DISTINGUISHING FEATURES The slender dwarf canes. The bright green leaves which stand erect and stiff, and also the bright yellow colour of the same when the older ones wither during the winter months.

CANES From 2 to $2\frac{1}{2}$ feet in height by $\frac{1}{4}$ inch in diameter. Bright green at first, maturing to a dull purple.

NEW SHOOTS Appear from late March onwards. Very slender. Pale green at first, ageing to deep green tipped purple.

CANE SHEATHS Persistent. Up to $3\frac{1}{2}$ inches in length, often covering the internode. Dull straw coloured.

NODES Not prominent. From 2 to 4 inches apart.

BRANCHES Usually borne singly at each node, but occur occasionally in pairs. Very slender. Upright in habit. Purple.

LEAVES Up to 7 inches in length by $\frac{3}{4}$ inch wide. Leaf-stalk short, pale green in colour. Midrib fine, but palpable for three-quarters of its length from the base. Base bluntly rounded, and the tip short, but sharply tapered. Edged on one margin with fine bristles, the other margin only partially so. Tessellation very obvious. Four to five pairs of secondary veins. Bright green on the upper surface, dull grey green on the lower.

PROPAGATION Division of the clump. Rhizome cuttings.

ROOTSTOCK Running—but though classed as such, the clump at Pitt White has not been observed to do so.

Arundinaria graminea
Syn. *Arundinaria hindsii* var. *graminea*
    Bambusa graminea
    Pleioblastus gramineus
Country of origin: Japan
Native name: Tai-min-chiku

A. graminea (grass-like) possesses the narrowest leaves in proportion to the length of the canes of all the medium-tall hardy bamboos. It cannot be suggested as a species for the smaller garden owing to the invasive qualities of its rootstocks. It can still, however, be used to good account for screening purposes, and as a shelter plant. It is one of the few hardy bamboos which prefer to grow in the shade. The mature canes, when properly ripened, are ideal for garden use.

Originally thought by Bean to be a variety of *A. hindsii*, but this has since been disproved by contemporary botanists.

A native of the Ryuku Islands of Japan, it is believed to have been introduced into England by Messrs. Veitch in 1877.

It has flowered on many occasions, and has done so as recently as 1965 at Pitt White. Seed obtained from the flowering was sown and germinated freely.

DISTINGUISHING FEATURES The long, bright green, distinctly grass-like leaves which are very numerous on the many branchlets, and the persistent sheaths.

CANES 10 feet plus in height by $\frac{3}{4}$ inch in diameter. Thin walled. Pale green at first, ageing to deep green, and maturing to a dull yellowish green.

NEW SHOOTS Appear from late May onwards. Pale green with purple tips.

CANE SHEATHS Persistent. Dull straw coloured.

NODES Rather prominent. Average 6 inches apart.

BRANCHES Usually borne in pairs or threes at first, then numerous at every node, especially at the summit of the cane. Four to six leaves per branchlet.

LEAVES Up to 10 inches in length by ½ inch wide. Shortly wedge-shaped at the base, and sharply tapered at the tip. Edged on both margins with fine bristles, particularly near the tip. Leaf-stalk short. Midrib fine but palpable. Tessellation fine but clear. Four to five pairs of secondary veins. Bright green on the upper surface, duller matt green on the lower.

PROPAGATION Division of the clump. Rhizome cuttings. Cane layering.

ROOTSTOCK Running. Fairly invasive and soon forms dense clumps.

Arundinaria hindsii

Syn. *Bambusa erecta*
     Pleioblastus hindsii
     Thamnocalamus hindsii
Country of origin: China
Native name: Kanzan-chiku

Originally a native of China, *A. hindsii* has been grown for centuries in Japan. It is somewhat similar to, and is often confused with, *A. graminea*, but the foliage of *A. hindsii* is always much coarser and wider than that of *A. graminea*.

It is perfectly hardy, can be grown easily in all types of soil, and does well under deep overhead shade. It can also be used as a hedging species. The canes of the mature clumps are useful in the garden.

According to Camus, this species flowered in France in 1912, and it was introduced into England *circa* 1875.

DISTINGUISHING FEATURES The olive-green, thin-walled canes. The ragged portions of the cane sheaths which often remain attached to the nodes on the older branches. The clustered, top-heavy branch formation.

CANES Up to 12 feet in height by 1 inch in diameter. Thin walled. Brilliant deep green at first, and covered with a waxy bluish-white bloom. Matures to a dull olive-green shade.

NEW SHOOTS Appear from April onwards. Pale green tipped with purple. Often mottled with purple blotches.

CANE SHEATHS Partially persistent. May remain attached to the base of the node for several seasons. Short, do not cover the internode.

NODES Not prominent. Can be as far apart as 10 inches.

BRANCHES From three to five in the first season, then numerous branches in succeeding seasons. Erect in habit. Branches form heavy clusters at the summit, giving the clump a top-heavy appearance. Four to five leaves per branchlet.

LEAVES Very variable in size, but can be as long as 9 inches by 1 inch wide. Stand fairly erect. Wedge-shaped at the base, and sharply tapered at the tip to end in a fine long tail. Thick and coarse textured. Midrib fairly prominent and palpable. Edged with fine bristles on one margin, partially so on the other. Tessellation clear. Leaf-stalk medium long. Four to six pairs of secondary veins. Dark green in colour on the upper surface, dull greyish green on the underside.

PROPAGATION Division of the clump. Rhizome cuttings. Cane layering.

ROOTSTOCK Running. Fairly invasive, and will travel freely if allowed to do so under good cultural conditions.

Arundinaria hookeriana
Country of origin: India
Native name: Praong

A. hookeriana is not quite hardy, and usually requires some greenhouse shelter. It can be grown outdoors only in very favoured areas. As a tub plant in a conservatory it is highly valued for its wealth of delicate foliage and the uniqueness of its colouring. The golden cane colour is not so striking as that of *P. castillonis*, but it is much more graceful in habit. A native of India, *A. hookeriana* grows at altitudes of between 4,000 and 6,500 feet above sea-level in the states of Sikkim and West Bhutan. In its native habitat this bamboo, in common with most of the 'high-altitude' bamboos, grows separate leaf- and flower-bearing canes every season, but when grown in temperate countries its habits change; the canes do not die down every year—they become sturdy and live for many years. Camus in his *Les Bambusees* writes of this species as having flowered in Sikkim, Bhutan, and in several parts of Europe in the 1845, 1885, and 1892/3 seasons.

DISTINGUISHING FEATURES The mature bright golden canes with their pink and green striations are instantly recognizable. The absence of visible tessellation in the foliage. The ragged portions of the cane sheaths which remain attached to the node.

CANES 15 to 20 feet in height by ½ to 1½ inches in diameter. Thin walled. Bluish green in colour at first, coated with a white scurf when young. Mature to golden yellow with thin green striations. In a sunny position the mature canes often show a deep pink colouring.

NEW SHOOTS Appear from late May onwards. Rose pink, quickly developing into a dull bluish-green shade.

CANE SHEATHS Fairly persistent. The basal portion may remain attached to the node long after the remainder of the sheath has withered away. Striated and papery in texture. Narrow to an abrupt point at the tip. Dull straw coloured. Sheath blade fringed with fine hairs.

NODES Not prominent, usually just a thin straw-coloured ring. Average 8 inches apart. On the mature canes there is often a dark bluish band of colour on the upper portion of the node, and this may also be striped with faint green and yellow streaks.

BRANCHES Numerous at every node.

LEAVES Very variable in size, ranging from 3 to 12 inches in length by ½ to 1½ inches wide. Smooth on the upper surface, the underside bears fine white hairs at the base of the midrib. Short and sharply tapered at the tip, narrowing abruptly to a blunt base. Leaf-stalk fairly short and thin. Edged with fine bristles on both margins. No visible tessellation, but carries numerous translucent glands. Seven to eight pairs of secondary veins. Pale bluish green on the upper surface, matt sea green on the underside.

PROPAGATION Division of the clump. Basal cane cuttings.

ROOTSTOCK Caespitose. Increases slowly. Makes an excellent tub plant.

Arundinaria humilis
Syn. *Arundinaria fortunei* var. *viridis*
    Bambusa nagashima
    Nipponocalamus nagashima
    Pleioblastus nagashima
    Pleioblastus humilis
Country of origin: Japan
Native name: Hirouzasa

Named by A. B. Freeman-Mitford as *humilis* (of low growth), it comes from the Honshu and Kyushu districts of Japan. Sometimes known as the green form of *A. fortunei*, to which, incidentally, it is no relation. Bean stated that there was no discernible difference between *A. humilis* and *B. nagashima*, and gave the latter as a synonym. I see no reason to doubt his statement.

A. humilis is a pretty semi-dwarf bamboo, but owing to the thinness of its canes, it is often cut to the ground during very cold weather. It invariably recovers to form a green carpet before the following summer is very old. As a quick cover for waste ground, for clothing banks, or stabilizing screes, this bamboo is excellent.

DISTINGUISHING FEATURES The length of the thin branches in proportion to the height of the semi-dwarf canes. The long pale green leaves. Can be differentiated from *A. vagans* by the paler leaf colour, the thinner canes and the length of the slender branches. Slightly downy on the underside of the leaf, whilst the leaves of *A. vagans* are downy on both surfaces.

CANES From 2 to 6 feet in height by $\frac{1}{4}$ inch in diameter. Dark green.

NEW SHOOTS Appear from May onwards. Dull purple.

CANE SHEATHS Not persistent. Purplish in colour at first, maturing to a drab yellow shade.

NODES Not prominent. Up to 5 inches apart.

BRANCHES Borne in pairs or threes at each node. Long in relation to the height of the cane, often 3 feet long.

LEAVES Up to 8 inches in length by $\frac{3}{4}$ inch wide. Slightly downy on the underside. Rounded at the base, and slenderly pointed at the tip. Midrib fairly prominent. Edged with fine hairs on one margin, partially so on the other. Four to six pairs of secondary veins which are also fairly prominent. Pale green on the upper surface, dull matt pale green on the lower.

PROPAGATION Division of the clump. Rhizome cuttings.

ROOTSTOCK Running. Can become troublesome in good growing conditions.

Arundinaria japonica
Syn. *Arundinaria metake*
    Bambusa japonica
    Bambusa metake
    Pseudosasa japonica
    Sasa japonica
    Yadakeya japonica
Countries of origin: China, Japan, Korea
Native names: Metake; Yadake

Sometimes known as the 'arrow' bamboo, *A. japonica* must be classed as the hardiest of the Arundinaria, and the most widely planted. It can be seen growing freely in most of the temperate countries. For hedging and screening purposes this species is without equal, and the mature canes, when ripened properly, are eminently suitable for garden purposes, even though they are fairly thin walled. It also makes a fine tub plant.

Known in Japan as the Yadake or 'female' bamboo, but why this should be so is a mystery, as there is no botanical reason to justify this wording. The native name Metake should not be confused with Medake (*A. simonii*) and Madake (*P. quilioi*).

Introduced into France by von Siebold in 1850, and from thence to England a few years later. It has been under cultivation in the United States of America for at least half a century.

Clumps of *A. japonica* have flowered in many areas in temperate countries. Camus records it as having flowered in Japan in 1877, 1886, and in 1898. It flowered in Paris and in Algeria at roughly the same time. Information received from the State Nikita Botanic Gardens at Yalta reveal that this species has flowered on many occasions in recent years. In spite of its prolificity of flowering, comparatively little seed was obtained. It was also revealed that very few of the clumps died out after the flowering, and most continued to grow normal canes.

DISTINGUISHING FEATURES The broad leaves in comparison to the height of the canes. The very persistent cane and branch sheaths. The oddly coloured strip on the lower side of the leaf surface. The slowly spreading clumps of the semi-arching canes.

CANES Up to 20 feet in height by 1 inch in diameter. Thin walled. Erect in habit when young, arching with the weight of foliage and branches when older. Olive green at first, maturing to a dull matt green shade.

NEW SHOOTS Appear from late April onwards. Deep glossy green edged with purple. Hairy.

CANE SHEATHS Very persistent, often clasping the extending branches even after the third season. Long, sometimes covering the internode. Very coarse in texture. Covered in the first season with close short hairs which drop off during the second year. Has a long awl-shaped tongue at the tip. The topmost sheaths often develop into true leaves, particularly on the fresh canes, and these 'sheaths' are invariably much larger than the true branch leaves. Green in colour at first, quickly turning to a drab straw shade.

NODES Not prominent. Up to 10 inches apart.

BRANCHES Usually borne singly, occasionally in twos or threes. Long.

LEAVES Up to 12 inches in length by 1¾ to 2 inches wide. Base bluntly rounded, occasionally wedge-shaped, and the tongue long and tapered at the tip. Leaf-stalk long, wide, and yellow in colour. The mid-rib is yellow and very conspicuous, being palpable for three-quarters of its length from the base. The lower side of the leaf is felted with fine white hairs, with the exception of a strip which is coloured as the surface. Very conspicuously tessellated. Edged with fine bristles on one margin and at the base and the tip of the other. Six to ten pairs of secondary veins. Dark glossy green on the surface, dull greyish green on the underside, with the exception of the oddly coloured strip.

PROPAGATION Division of the clump. Rhizome cuttings. Cane layering.

ROOTSTOCK Although classed as running, it rarely travels far afield, unless conditions are exceptionally favourable, and even then it seldom becomes invasive.

Arundinaria laydekeri
Syn. *Arundinaria chino* cv. 'Laydekeri'
  Pleioblastus chino var. *laydekeri*
Country of origin: Japan
Native name: Kinjo-chiku

Similar in growth and habit to type *A. chino*, and often has dull irregular yellow variegations in the foliage. These variegations are very erratic, however, and cannot be relied on. Like its type, it is rather a dull and uninteresting bamboo, and cannot be recommended as a good garden species.

Arundinaria macrosperma
Syn. *Arundinaria gigantea*
  Bambusa newmanii
Country of origin: The United States of America
Native names: Southerncane, Canebreak

A. macrosperma is the tallest of the few species which are native to the North American continent. It is found mainly in the southern and south-eastern states which border on the Gulf of Mexico. In these areas this bamboo grows, or used to grow, in vast thickets on the verges of the many river estuaries. These thickets were reputed to have been used as places of concealment by runaway slaves making their way north to freedom during the Civil War years.

A. macrosperma (bearing long seeds) is slightly tender and must be planted in a sheltered position. It flowers on the old stems and on the leafless branches in its native habitat, but has not been recorded as having done so in any area outside its homeland.

DISTINGUISHING FEATURES The dull yellowish colour of the canes. The thinness of the walls in comparison to the size of the canes. The persistent sheaths, and the short, slender, numerous bundles of branches.

CANES 20 feet plus in height by up to $1\frac{1}{4}$ inches in diameter. Thin walled. Dull greenish yellow.

NEW SHOOTS Appear from late May onwards. Deep purple.

CANE SHEATHS Very persistent. Fringed at the tip with coarse hairs. Coarse and striated. Purplish in colour at first, fading to a dull straw shade.

NODES Fairly prominent. Up to 2 feet apart.

BRANCHES Numerous at every joint. Slender, short, and divergent. Often grow in dense bundles. Dull yellow in colour.

LEAVES Variable in size, can be as large as 14 inches in length by $1\frac{1}{2}$ inches wide on the older canes, but usually much smaller on the younger. Smooth on the upper surface, downy on the underside. Bluntly rounded at the base, short but sharply tapered at the tip. Edged with fine bristles on both margins. Leaf-stalk medium length. Tessellation very obvious. Up to fifteen pairs of secondary veins. Soft light green on the upper surface, light matt green on the lower.

PROPAGATION Division of the clump. Rhizome cuttings. Basal cane cuttings.

ROOTSTOCK Running. Free moving under good growing conditions, but is seldom aggressively invasive.

Arundinaria marmorea
Syn. *Arundinaria kokantsik*
    Arundinaria matsumurae
    Bambusa marmorea
    Chimonobambusa marmorea
Country of origin: Japan
Native name: Kan-chiku

A. marmorea

A. marmorea is an attractive, hardy, semi-dwarf bamboo which comes from the Honshu district of Japan.

Given the name *marmorea* or marble-like by M. Marliac, and introduced by him into France in 1889. It was brought over to England and Ireland shortly afterwards.

The rhizomes can be extremely active, and ample space must be allowed if this bamboo is to be seen at its best. In open spaces it is ideal for covering or screening, and always looks best when grown in a mass.

Flowered at Garstang in 1931, and at Kew in 1954.

DISTINGUISHING FEATURES The unusual colouring of the new shoots and cane sheaths. The canes though slight in diameter are very thick walled. The attractive arrangement of the first year branches, and the bright green leaves with their constricted tongue at the tip.

CANES Up to 6 feet in height by $\frac{1}{2}$ inch in diameter. Very thick walled. Beige green at first, maturing to a dull purple or deep purple when grown in a sunny position.

NEW SHOOTS Appear from late February onwards. Pale green mottled brown and silvery white. Tipped and striped with pink.

CANE SHEATHS Persistent. Fringed at the base with white hairs. Purplish at first and dotted with pinkish grey mottling. Ages to a dull silver grey shade.

NODES Rather prominent. Up to 6 inches apart.

BRANCHES Normally borne in threes, one short and two long at each node. Occasionally there may be up to five branches, but there is always one branch which dominates the group. The early branch formation is arranged in a herringbone pattern, and growth is tufted in the following year.

LEAVES Up to 6 inches in length by $\frac{1}{2}$ inch wide. The base is bluntly pointed, and the tip has a short slender taper and ends in a tiny tongue. Edged on one margin with fine bristles, only partially so on the other. Midrib prominent. Tessellation fair. Four to six pairs of secondary veins. Bright green on the upper surface, matt greyish green on the lower.

PROPAGATION Division of the clump. Rhizome cuttings.

ROOTSTOCK Running. Slow to start moving, but once established can become very active. Should make a good tub plant.

Arundinaria murielae

Syn. *Sinarundinaria murielae*

Country of origin: China

. *P. nigra*, Pitt White, Uplyme, Devon

11.
Bamboos at the Botanic Gardens, Dublin. By courtesy of the Irish Tourist Board

2. *A. japonica*, growing by the River Lym, Pitt White, Uplyme, Devon

13. *A. niitakayamensis* (Pitt White clone) at Pitt White, Uplyme, Devon. The tall canes tower over the author's son Alan (foreground)

A. murielae is a slender, graceful, and arching bamboo, somewhat resembling *A. nitida*, with which it has often been confused. The main difference lies in the colour of the canes and sheaths. On *A. nitida* they are purple, while on *A. murielae* they are bright pea green.

The country of origin of this species is China, where it grows at an altitude of 10,000 feet above sea-level in Hupeh Province. Though seemingly delicate, this classical bamboo is very hardy and can be seen growing freely in many parts of Europe, from the North of Scotland to the South of France.

Discovered by Wilson in 1907. In the same year the specific name *murielae* was given in honour of his daughter Muriel. First propagated in Europe in the Royal Botanic Gardens at Kew in 1913. The canes, though thin and seemingly fragile, can when ripened properly be utilized as garden plant supports.

DISTINGUISHING FEATURES A typical Arundinaria with slender canes of medium height and tessellated leaves, of the group which includes *A. nitida*, *A. hindsii*, *A. angustifolia*, and *A. graminea*.

From *A. nitida* it is distinguished by the pale green of its stems and shoots, instead of brownish purple, and by its new shoots acquiring branches at an earlier date, also by the pale pea-green colouring of the leaves.

The caespitose rootstock ensures that the clump is compact and spreads outwards at the circumferences slowly, whilst *A. graminea* and *A. hindsii* each have eager running rootstocks and form clumps with ragged edges. The canes of *A. murielae* do not form the finely twigged and bushy-tufted tops of *A. graminea* and *A. hindsii*, neither are they as wandering as those of *A. angustifolia*. The canes of this species are not as thin walled as the other three species.

CANES 14 feet plus in height by $\frac{1}{2}$ inch in diameter. Bright green during the first season, ageing to a deeper green, and maturing to a dull yellow shade.

NEW SHOOTS Appear from late April onwards. Smooth. Soft pale green with faint purple lines on the margins.

CANE SHEATHS Not persistent, although the late-emerging canes may retain their sheaths until the following spring. Roughly 6 inches in length. The sheath is smooth, except where it joins the cane, and there it is bristled with fine hairs. Rounded at the tip, and ends in an awl-shaped tongue. Pale green to cream at first, but quickly fades to a dull straw shade.

NODES Not prominent. Roughly 6 inches apart. Dusted during the first months with a white deposit.

BRANCHES Usually three to four during the first season. Numerous branches and twigs when mature which make the cane arch gracefully outwards.

LEAVES Up to 3 or 4 inches in length by $\frac{1}{2}$ to $\frac{3}{4}$ inch wide. Bluntly rounded at the base, and tapering to a short slender tip. Edged with fine bristles on one margin, partially so on the other. Leaf-stalk fairly long for the size of the leaf. Midrib fine. Tessellation very good. Three to four pairs of secondary veins. Light rich pea green on the upper surface, duller pale green on the lower.

PROPAGATION Division of the clump. Rhizome cuttings.

ROOTSTOCK Caespitose. Creeps only slowly. Makes an exceptionally good specimen plant.

Arundinaria niitakayamensis (Pitt White clone)

Syn. *Indocalamus niitakayamensis*
    Sasa niitakayamensis
    Yushiania niitakayamensis
Countries of origin: Formosa, The Philippines
Native name: Utod (Philippines)

A. niitakayamensis, an Asiatic species of Arundinaria, was first collected in 1905 on Mount Morrison in the island of Formosa (Taiwan), at altitudes of between 7,000 and 10,000 feet above sea-level, and named by B. Hayata in the *Botanical Magazine* of Tokyo (vol. 21, p. 49, 1907). A few years later it was discovered growing on the upper slopes of several mountains in the Philippine island of Luzon, Mt. Pulog, Mt. Ugo, etc.) and on Mindosa (Mt. Halcon). In the Philippines it is very abundant, forming dense thickets along the upper borders of the forests. On the upper limits it grows as a dwarf no taller than a few inches in height, but on the lower slopes the canes can be several feet tall.

The Pitt White clone produces canes of far greater dimensions, and I have measured canes of 32 feet in length. The rate of growth is phenomenal, often as much as 6 inches in twenty-four hours. The original material from which the clump at Pitt White was grown came from Perry's Hardy Plant Farm at Enfield in Middlesex in 1940, but the stock which remained at the nursery was obliterated when the ground was cleared for the growing of foodstuffs at the start of the 1939–45 war. It is believed that the clump at Pitt White was the only survivor. Propagation material from the clump has now been distributed to Perry's Hardy Plant Farm, the Royal Botanic Gardens at Kew, the Horticulture Research Station at Camborne, Cornwall, the Scottish National Gardens property at Inverewe, Tresco Abbey Gardens in the Scilly Isles, and to the National Botanic Gardens in Dublin. Intrigued by the difference in habit and tallness of the canes from the original Japanese description, material from the clump at Pitt White was sent to Tokyo on two occasions through the courteous co-operation of the Director of the Royal Botanic Gardens at Kew. On the first occasion the material was examined by Nakai himself, and the second by Dr. Muroi of the Fuji Bamboo Gardens at Gotemba. Each identified the plant as *A. niitakayamensis* (Hayata)—Nakai in spite of the thickness and length of the canes. The Pitt White clump has now been officially classified by the botanical authorities as *A. niitakayamensis* (Pitt White clone). The original *A. niitakayamensis* in its wild state flowered in Formosa in 1905, and in the Philippines in 1909, but little has been written on the flowering characteristics.

It was transferred to the Sasa group by E. G. Camus, who separated the Philippine plant as a distinct variety, *Saza niitakayamensis* (Hayata) E. G. Camus var. *microcarpa*, in his monograph on the bamboo (*Les Bambusees*, p. 24, 1913).

DISTINGUISHING FEATURES The smallness of the leaves in comparison to the very tall slender canes, and the drooping habit (as seen in Chinese and Japanese prints) when the tip of the cane bends downwards in the second season of growth under the weight of the myriads of delicate leaves borne on the multiple twigs.

CANES 30 feet plus in height by up to $1\frac{1}{2}$ inches in diameter. Fairly thin walled for the size of the cane. Upright in habit during the first season, then cascading during the following seasons. Deep glossy green at first, ageing to a yellowish green, and maturing to a dull brown shade.

NEW SHOOTS Appear from late May onwards. Smooth. Deep glossy green shaded pale purple at the tip. Edible.

CANE SHEATHS Not persistent on the early canes, but are retained on the late summer shoots. Striated and thick textured. Tongue long and narrow with a fringe of hairs at the tip.

NODES Not prominent. Average 2 feet apart, with a fine, thin, sharply defined ring marking the lower limits of the node. On the new canes there is a fairly thick coating of whitish bloom on the underside of the node, but this disperses quickly.

BRANCHES The larger canes do not carry branches on their lower six feet. Those at the tip are the first to appear, the lower ones opening in succession. Slender and arching. Three to four branches per node in the first season, multiple with numerous twigs thereafter.

LEAVES Average 3 inches in length by ½ inch wide; occasionally leaves may appear which are up to 6 inches in length, but these only appear on the younger canes. Midrib fine. The base is bluntly wedge-shaped, and the tip long and finely tapered. Leaf-stalk short. One margin fringed with fine bristles. Tessellation fair. Four to five pairs of secondary veins. Mid-green on the upper surface, dull greyish green on the underside.

PROPAGATION Basal cane cuttings. Division of the clump.

ROOTSTOCK Running. Of an uncommon type, *A. anceps* being the only species with a similar rootstock. Essentially caespitose, but with long running axes branching off from the short rhizomes of the standing culms. They are segmented and sheathed, but are devoid of any kind of lateral buds, and each ends in a short typical rhizome from the tip of which a new cane springs ready to multiply itself by tillering. They act as barren and rootless connecting links from the side of one rhizome to the base of another; a kind of grossly elongated rhizome neck. The resultant growth above ground is a compact clump surrounded at a short distance by scattered groups of new canes.

Arundinaria nitida
Syn. *Sinarundinaria nitida*
Country of origin: China

A. nitida is without doubt the most graceful of all the Arundinaria family, and, in fact, in some countries it is known as the 'Queen of the Arundinaria'. The leaves dislike the direct heat of the sun, and the margins will bend over towards the midrib, forming a channel, at the first touch of brilliant sunshine. As soon as the clouds again cover the sun the leaves will once more open out to their graceful normal shapes. It pays therefore to grow this dainty bamboo in a position where there is some partial overhead cover.

Although seemingly delicate, *A. nitida* is very hardy, and will only suffer a bit of leaf scorch in a very hard winter. Named as *nitida* (shining or lustrous) by A. B. Freeman-Mitford, which is an apt name for an outstanding plant.

It is often confused with that other deceptively delicate-looking bamboo *A. murielae*, but the essential difference lies in the cane and

A. nitida

sheath colouring. These are green on *A. murielae* and purple on *A. nitida*. There are, however, many clumps of *A. nitida* in existence which are of different clonal ancestries. These may bear canes with diverse shades of purple, from light to dark. The pure type has canes of deep rich purple.

A. nitida was introduced into England by Messrs. Veitch, who raised the plants from seed sent to them by Dr. Regel, the then Director of the Imperial Botanic Gardens at St. Petersburg, now Leningrad, in the U.S.S.R., in 1889. The original seed was believed to have been collected by an M. Berezovski in the South Kansu province of China in the early 1880s.

It has never been recorded as having flowered outside its native habitat, and the Chinese believe that this species flowers only once in every hundred years.

DISTINGUISHING FEATURES The very slender purple canes are densely packed into a closely circumscribed clump. Their lower halves are bare, and rise almost vertically, but at the higher levels, the wealth of foamy foliage bends them well outwards in a filmy mass. The individual leafed branchlets are so fine that the foliage gives an impression of floating in the air. The thin whiplike new canes sway above the mass of foliage below them.

CANES 12 to 15 feet in height by up to $\frac{3}{4}$ inch in diameter. Whiplike and branchless in the first season. Have a heavy coating of bloom, bluish grey in colour, which may remain for several years when the plant is growing in a very shady spot, but disperses before the end of the second season if the planting position is more open. Greenish purple in the early part of the first season, quickly turning to deep purple. When grown under heavy overhead cover the cane colour may remain greenish purple.

NEW SHOOTS Appear from late May onwards. Pale green mottled with pink, cream and brown. Coated with fine hairs.

CANE SHEATHS Persistent. Hug the cane tightly during the first season of growth until the spring of the second. Hairy and long, often covering the internode. Thin textured. Pale purple during the first season, ageing to a dull straw shade before being discarded.

NODES Not prominent. Roughly 6 inches apart. The ring is thin, creamy white in colour, and quite distinct.

BRANCHES Branchless during the first season, four to five branches per node in the second, then numerous branches and twigs in the following seasons. Dull purple coloured.

LEAVES Small, averaging 3½ inches in length by ½ inch wide, papery thin. Leaf-stalk short and very fine. Midrib slender. Bluntly rounded at the base, and abruptly short tapered at the tip. Edged with fine bristles on both margins, especially towards the tip. Tessellation visible, but indistinct to the naked eye. Three to four pairs of secondary veins. Brilliant mid-green on the upper surface, matt sea green on the lower.

PROPAGATION Division of the clump. Rhizome cuttings. Basal cane cuttings.

ROOTSTOCK Caespitose. Very slow moving. Makes an exceptionally good specimen plant and tub subject.

Arundinaria pumila

Syn. *Bambusa pumila*
    Nipponocalamus pumilus
    Pleioblastus pumilus
    Sasa pumila
Country of origin: Japan
Native name: Sudore-yoshi

A. pumila is an extremely hardy dwarf bamboo, second only in small-ness to *A. pygmaea*. A native of the Honshu, Shikaku, and Kyushu districts of Japan, it can be used to some advantage for clothing banks, or as a soil-stabilizing medium in loose soil areas. It was named *pumila* (little) by A. B. Freeman-Mitford when it was introduced into England in the latter part of the nineteenth century.

DISTINGUISHING FEATURES The very dwarf habit of growth, the hairy leaf surfaces, and the smaller and more sharply tapered leaves, plus the ring of hairs at the base of the sheath help to differentiate it from *A. vagans*, with which it has often been confused.

CANES Up to 2½ feet in height by ¼ inch in diameter. Very slender. Bloomed under the nodes with a thick white deposit. Growth a little flattened on the upper parts. Dull purple in colour.

NEW SHOOTS Appear from early May onwards. Dull green coloured.

CANE SHEATHS Fairly persistent. The sheaths borne higher up the cane have a conspicuous ring of hairs at the base. Purple tinged at first, fading to a dull straw shade.

NODES Not prominent. Up to 6 inches apart. Bloomed on the under-side.

BRANCHES Usually borne singly, but occasionally in pairs. Five to seven leaves per branchlet.

LEAVES Up to 7 inches in length by ¼ to ¾ inch wide. Narrowly lanceolate. Rounded at the base, and narrow to an abrupt point at the tip. Covered with minute close hairs on both surfaces. Edged on both margins with fine bristles. Tessellation clear. Four to five pairs of secondary veins. Intense dark green on the upper surface, dull matt green on the lower.

PROPAGATION Division of the clump. Rhizome cuttings.

ROOTSTOCK Running. Very free moving; can become troublesome if it is not kept under strict control.

Arundinaria pygmaea

The term 'pygmaea' has in the past been applied to several species. Camus describes *Sasa variegata* as having a variety 'viridis' which occurs in two forms, 'pubescent' with downy hairs on each surface of the leaf, and 'glabrous', with each leaf surface smooth and devoid of hairs. The 'pubescent' form is the *A. pygmaea* of Mitford, the 'glabrous' form is *A. variabilis* var. *pygmaea* (Makino), syn. *B. pygmaea* of the Japanese botanists, but not Miguel. *B. pygmaea* (Miguel) is probably Mitford's form, or possibly *B. disticha*.

Arundinaria pygmaea (pubescent form)

Syn. *Bambusa pygmaea*
    Pleioblastus pubescens
    Pleioblastus pygmaeus
    Sasa pygmaea
    Sasa variegata var. *pygmaea*
Country of origin: Japan
Native names: Ke-nezasa; Ke-oroshima-chiku

This tiny bamboo comes from the Honshu, Kyushu, and Shikaku provinces of Japan, where it grows freely and covers the forest floors with masses of tiny canes. It is so small that the American botanist Young, who unlike many botanists accepts it as a good species of Sasa, stated in the *Nat. Hort. Mag.*, 180, 1945, 'because of the slenderness of the canes, it may even be possible to mow it.'

A. pygmaea in either of its forms cannot be recommended for the smaller garden owing to the rapidity with which its rhizomes spread, but it can be utilized as a quick ground-cover type, and for planting as a soil-stabilization agent in loose soil and river-bank areas.

DISTINGUISHING FEATURES Easily distinguished by its very dwarf habit of growth. It is, in fact, the dwarfest of all the hardy bamboos. Can be distinguished from the 'glabrous' form by the fine hairs on the leaf surfaces, but it it is often confused with *A. vagans* and even *S. veitchii*, with the former when it is drawn by lack of light, making the canes longer than normal, and with the latter by the occasional broad leaves which sometimes appear and by the identical leaf-tip and marginal withering. There is, however, a slight difference in the withering, as the midrib section of the *pygmaea* leaf is often seen to be decayed.

CANES Up to 10 inches in height by $\frac{1}{16}$ inch in diameter. Round, slender, and solid. Flattened at the tip. Bright green, becoming purplish towards the tip.

NEW SHOOTS Appear from late April onwards. Bright green.

CANE SHEATHS Fairly persistent. Dull green, ageing to a dull muddy straw. The blade is small but broad, bright green, and covered with minute hairs.

NODES Prominent for the size of the cane. Fringed with minute bristles. Have a waxy band on the underside. Purple.

BRANCHES Solitary as a rule, but occasionally in pairs. Long in proportion to the size of the cane.

LEAVES Up to 5 inches in length by $\frac{3}{4}$ inch wide, occasionally wider and longer on the terminal extremities. Base rounded, and pinched in some $\frac{1}{2}$ inch from the tip to end in a short sharp point. Edged with minute bristles on both margins. Leaf-stalk very short. Midrib fine. Both surfaces of the leaf are covered with fine hairs, the ones on the upper surface being much coarser than those on the lower. Tessellation only fair. Three to five pairs of secondary veins. Brilliant green on the upper surface, dull silvery green on the lower. Tend to wither on the tips and margins, and also in the midrib section.

PROPAGATION Division of the clump. Rhizome cuttings.

ROOTSTOCK Running. When established it can become an absolute pest if it is allowed to stray outside its allotted position.

Arundinaria pygmaea (glabrous form)
Syn. *Arundinaria variabilis* var. *pygmaea*
    Bambusa pygmoea
    Pleioblastus variegatus forma *glabra*
    Sasa pygmaea
    Sasa variegata var. *viridis* forma *glabra*
Native name: Ne-zasa

Occurs in two sub-varieties, which are botanically akin to one another, lacking the hairs on the leaf surfaces. There is little noticeable difference between them to the casual eye.

1. Sub-variety *Akebono*

CANES Slender with rather long branches.

LEAVES Close together. Narrowly lanceolate or linear lanceolate. Narrowed or obtuse at the base, and sharply pointed at the tip. Thin and papery in texture. Smooth on both surfaces. Bright green.

2. Sub-variety *Tanake*

CANE SHEATHS Not fringed with hairs at the throat.

LEAVES Larger than the type, and either linear lanceolate or oblong lanceolate. Bright green.

Arundinaria racemosa
Country of origin: India

Originally a native of the Indian subcontinent, from the states of Sikkim and Bhutan, *A. racemosa* is found growing at altitudes of between 6,000 and 12,000 feet above sea-level. It is used extensively in these areas to provide cattle fodder, and for many domestic purposes. It is perfectly hardy, and at first glance may be confused with *A. spathiflora*, as it has the same habit of growth, and somewhat similar though darker foliage. It is, however, much more robust than that species.

A. racemosa was named by Munro. His original description, however, was based on two different species, the dwarf highland *A. racemosa* and a taller lowland species, *A. malig*. The muddle was sorted out by Gamble, and the description corrected in the *Kew Bull.*, 1912, p. 135.

DISTINGUISHING FEATURES The brownish green canes, and the rough texture of the internodal surfaces, which is not only readily noticeable, but can be felt quite distinctly.

CANES Up to 15 feet in height by $\frac{1}{2}$ to 2 inches in diameter. Extremely rough to the touch between the internodes. Brownish green at first, maturing to dull brown.

NEW SHOOTS Appear from late May onwards. Hairy. Pale fawn, streaked and blotched with pale purple.

CANE SHEATHS Persistent. Coarsely striated. Coated with smooth hairs when young, smooth later. Blade and margins fringed with fine hairs.

NODES Not prominent.

BRANCHES Normally borne in pairs or threes at each node.

LEAVES Up to 6 inches in length by $\frac{1}{2}$ inch wide. Rounded at the base, and tapered to a fine point at the tip. Leaf-stalk short. Midrib fairly conspicuous. Edged with fine bristles on both margins. Hairy on the lower surface. Tessellation conspicuous. Three to five pairs of secondary veins. Medium deep green on the upper surface, dull matt green on the lower.

PROPAGATION Division of the clumps. Rhizome cuttings.

ROOTSTOCK Running. Though classed as such, it seldom becomes invasive.

Arundinaria simonii

Syn. *Bambusa simonii*
    Bambusa viridistriata
    Nipponocalamus simonii
    Pleioblastus simonii
Countries of origin: China; Japan
Native names: Medake; Kawa-take

A native of China and the Honshu, Shikaku, and Kyushu districts of Japan, *A. simonii*, besides being one of the taller bamboos in the Arundinaria, is a very hardy and vigorous grower.

Introduced originally into France in 1862 by M. Simon, the French Consul in China at the time, and named after him by M. Carriere. The mature canes, though fairly thin walled, are very tough, and are ideal for garden use when properly ripened.

A. simonii has flowered on numerous occasions in France, Japan, and in Belgium, and most of the *A. simonii* alive today are believed to be descendants of plants grown from seed obtained in 1903/4, when the species flowered profusely in many parts of the globe. As recently as the 1962/3 seasons it flowered at Yalta in the Crimea, but according to information received from the Curator, A. M. Kormilitzin, no viable seed was obtained, but the clumps themselves did not die out after the flowering, as fresh growth was observed soon after the dead canes were removed. At the time of writing several clumps at Pitt White are in bloom, and a quantity of seed has been obtained.

A. simonii is grown in many countries which have reasonable temperate climates, and in the United States of America, where it does particularly well, it is known as the 'Simon' bamboo. This species can be planted as an alternative to the commoner *A. japonica*, as it is very suitable for use as a hedging plant.

DISTINGUISHING FEATURES The tall olive-green, fairly thin-walled canes, the multiplicity of the branches and their singular steeple formation. The two quite distinct shades of colour on the underside of the leaf, similar to the colouring of *A. japonica*. This species is not unlike *A. fastuosa* in habit, though not as stately, and it also lacks the unique interior sheath colouring of that species.

CANES 20 feet plus in height by up to $1\frac{1}{2}$ inches in diameter. Erect habit of growth. Thin walled but tough. Dusted with a white deposit between the nodes in the first season. Deep olive green at first, ageing to a dull green shade.

NEW SHOOTS Appear from May onwards. Deep olive green lightly edged with purple on the margins. Edible.

CANE SHEATHS Very persistent, often staying on the canes until completely desiccated. Roughly 10 inches in length. Highly glazed on the interior surface. Dull green tinged with purple on the margins when young, quickly turning to a dull brown.

NODES Not prominent, just a thin straw-coloured ring. Can be as far as 14 inches apart.

BRANCHES Usually borne singly on the larger canes during the first season, then two per node, finally maturing to a dozen or more arranged in a steeple formation. The smaller-diameter canes may have a great number of branches in the first season, but one branch in the centre of the group always dominates the rest. Three to six leaves per branch.

LEAVES Can reach up to 12 inches in length by $1\frac{1}{4}$ inches wide on the older canes, but considerably smaller on the younger canes. Bluntly pointed at the base, and long tapered at the tip to end in a fine tail. The leaf-stalk is long, thin, and pale green. Edged with fine bristles on one margin, partially so on the other. Midrib prominent and palpable for three-quarters of its length from the base. Tessellation clear and prominent. Five to seven pairs of secondary veins. Vivid green on the upper surface, the underside is green on one side and greyish green on the other, especially towards the tip.

PROPAGATION Division of the clump. Basal cane cuttings. Cane layering.

ROOTSTOCK Running. Can never be classed as invasive.

A. simonii

Arundinaria spathiflora

Syn. *Thamnocalamus spathiflorus*

Country of origin: India

A. spathiflora is an extremely handsome and elegant bamboo from the states of Sikkim, Bhutan, and Nepal in the Himalayas, where it grows at altitudes of between 7,000 and 10,000 feet above sea-level.

It is not very hardy, and should only be grown in more sheltered areas away from cold winds. Neither does it like the full sun on its foliage, as in common with the other thin-leaved species the leaves will curl up at the edges when exposed to the full glare of the sun, but open again to their full width when shaded.

At one time in India this bamboo supplied the bulk of the raw material used in the manufacture of walking-sticks, umbrella handles, pipe stems, and numerous other articles.

A. spathiflora (flowers borne in spathes) was introduced into England in 1886.

According to Camus, this species flowered in India on numerous occasions between 1821 and 1893.

DISTINGUISHING FEATURES Delicate in appearance, *A. spathiflora* is easily recognized by the thinness of its leaves, and by the pinkish-purple colour of its canes. It is conspicuous also by the neat habit of growth, which is in close circumscribed clumps free from any excessive arching.

CANES Up to 15 feet in height by 1 inch in diameter. Erect in habit, with an inclination to zigzag. Dusted with a bluish-white bloom throughout the first season. Bright green at first, ripening to a pinkish purple shade on the exposed side and pale green on the other.

NEW SHOOTS Appear from late May onwards. Pale green, spotted with purple, and often streaked purple. Hairy.

CANE SHEATHS Not persistent. Roughly 7 inches in length. Rounded at the tip, and fringed with fine hairs at the base, ending in an awl-shaped tongue. Highly polished on the interior surface.

NODES Conspicuous. Have a distinct white ring on the lower part of the node. Stained brown on the upper surface.

BRANCHES Normally borne in pairs or threes at each node. Pale purplish pink in colour.

LEAVES Up to 6 inches in length by $\frac{1}{2}$ inch wide. Thin and papery in texture. Both margins are fringed with fine bristles. Base bluntly pointed, and tip is slender and sharp. Leaf-stalk short, pale purple in colour. Midrib very fine. Three to five pairs of secondary veins. Pale green on the upper surface, matt greyish green on the lower. Tessellation fine.

PROPAGATION Division of the clump. Basal cane cuttings.

ROOTSTOCK Caespitose. Slow moving, making it an excellent specimen plant for a sheltered spot, and a good tub plant.

Arundinaria tecta

Syn. *Arundinaria macrosperma suffruticosa*

Arundinaria macrosperma var. *tecta*

Country of origin: The United States of America

Native names: The switchcane; the smallcane

A. tecta is regarded by some authorities as being a variety of the other North American bamboo, *A. macrosperma*. This may or may not be so, but it is certainly a prolific grower in the southern and south-eastern states bordering on the Gulf of Mexico. It grows in vast tracts in the swampy river deltas.

As a garden bamboo, *A. tecta* is rather second rate, and although shrubby in appearance, it always looks sad and tattered during the winter months. Originally believed to be semi-hardy, it has since been found to be fairly cold tolerant, and will grow freely in gardens where the all-year-round temperature can be reasonably well maintained.

Its date of introduction into Europe is uncertain. Given the name *tecta* (covered) by Muhlenberg *circa* 1877.

The flowering features of this species are similar to *A. macrosperma* in that it flowers on the leafless shoots of the year, but is not on record as having done so outside the United States. The thin-walled short stubby canes are of little value in garden work, but the foliage and the succulent new shoots have been found to provide excellent grazing material for cattle in the 'tecta' growing areas in the United States.

DISTINGUISHING FEATURES The persistent sheaths. The upright habit of the branches, and the overall shabbiness of the semi-dwarf canes with their partly withered foliage during the winter months.

CANES Up to 6 feet in height by $\frac{1}{2}$ inch in diameter. Thin walled. Smooth. Mid-green, ageing to a dull olive-green shade.

NEW SHOOTS Appear from late May onwards. Dull green, edged purple.

CANE SHEATHS Persistent, often adhering to the canes until tattered. Fringed with fine hairs. Dull straw coloured.

NODES Not prominent. From 4 to 6 inches apart.

BRANCHES Usually borne in pairs or threes, but sometimes up to six on the higher limits. Upright in habit.

LEAVES Up to 10 inches in length by ¾ to 1 inch wide. Coarse in texture. Midrib prominent and palpable up to half-way from the base. Downy on the underside. Bluntly rounded at the base, and short tapered at the tip. Leaf-stalk fairly small. Tessellation good. Fringed on both margins with fine bristles. Four to five pairs of secondary veins. Bright green on the upper surface, dull green on the lower. The leaf margins and tips tend to wither during the cold winter months.

PROPAGATION Division of the clumps. Rhizome cuttings.

ROOTSTOCK Running. Fairly active and rambling in milder districts, but slow moving in colder areas.

Arundinaria tessellata
Country of origin: South Africa
Native name: Bergbamboes

This is the only indigenous bamboo so far recorded in South Africa. It grows prolifically in many areas of that country, from Table Mountain in the south to Basutoland and Natal in the north, at heights of between 5,000 and 8,000 feet above sea-level on the mountains of Kat Berg, Winter Berg, and Witte Berg. It is found growing in vast tracts in sheltered ravines and wet places. It is known by the Afrikaans name of 'bergbamboes', and one mountain is actually named Bamboesberg or 'mountain of the wild bamboo' after *A. tessellata*, which grows so freely there. This species has often been confused in the past, both when written about in bamboo literature and in gardens, with *Sasa tessellata* (syn. *S. ragamowski*). No two bamboos can be more dissimilar in growth or habit as these two species, and if one saw a clump of each growing side by side, there would be little doubt as to which was which. The confusion arose, no doubt, by the similarity of the names, but this is just another instance of the confused state of bamboo nomenclature at the present time.

The name *tessellata* (marked with small irregular squares) given to this species by Munro can be confusing, as the tessellation in the leaves of this species is no more prominent than it is in the leaves of most hardy bamboos. Some species, *A. japonica*, for instance, have leaves which have an even more distinct tessellation than the leaves of *A. tessellata*,

4. Well established clump of *A. nitida*

15. Screening hedge of *A. anceps*

so it is a mystery why Munro chose that particular name for this specific bamboo. The heights and widths of the canes and foliage given in the description below are the average measurement taken from clumps grown in England at the time of writing. In South Africa the canes grow from 5 to 10 feet tall in dry places, and from 15 to 20 feet in moist areas.

In South Africa the ripe canes were used at one time by the Zulu warriors to reinforce the framework of their hide-covered shields, and by the forest tribes as arrow shafts and spear handles. The canes of this species grown in the United Kingdom, though thin walled, can be used as plant supports, but they must be ripened thoroughly first of all.

DISTINGUISHING FEATURES *A. tessellata* is easily identified from other bamboos by the close crowding of the lower nodes, reminiscent of that other odd bamboo, *Chusquea culeou*, and it shows nearly the same bottle-brush effect in the branch growth. No other hardy bamboo, however, has the pure white sheaths this species possesses, and these alone are an infallible means of identification.

CANES Up to 12 feet in height by $\frac{1}{2}$ inch in diameter. Thin walled. Erect in habit. Pale green at first, maturing to a deeper green, ageing to deep purple. The purple is more intense when grown in an exposed position.

NEW SHOOTS Appear from late May onwards. Densely covered with fine white hairs at first, smooth later. Pale creamy green, often flushed with pale pink streaks.

CANE SHEATHS Very persistent. 3 to 5 inches in length. Hairy at first. Papery thin in texture. Highly glossy on the interior surface. Pure white during the first season, ageing to a dull cream shade. The branchlet sheaths are also persistent and coloured as the cane sheaths.

NODES Not prominent. The spacing varies considerably, from $1\frac{1}{2}$ to 8 inches.

BRANCHES Usually branchless during the first season, but numerous branches with plentiful twigs appear during the following seasons. Growth short and tufted. From three to four leaves per branchlet.

LEAVES From 2 to $5\frac{1}{2}$ inches in length by $\frac{1}{2}$ inch wide. The base is obtusely rounded, and tapers in the final quarter of its length to end in a slender point at the tip. Leaf-stalk medium long and hairy. Edged with fine bristles on one margin, partially so on the other. Tessellation fine but clear. Midrib fairly prominent. The underside of the leaf is slightly hairy towards the base. Four to five pairs of secondary veins. Mid-green in colour on the upper surface, matt green on the lower.

PROPAGATION Division of the clump. Rhizome cuttings.

ROOTSTOCK Running. Under cool conditions it cannot be called invasive.

Arundinaria Vagans
Syn. *Pleioblastus viridistriatus vagans*
Country of origin: Japan

Known commonly as the 'dwarf' bamboo, *A. vagans* was brought from Japan to England in 1892.

Owing to the rapidity with which its rhizomes spread, this bamboo cannot be recommended for cultivation in the smaller garden, and when grown in the larger garden it should be relegated to a position where it can have plenty of freedom. In such a position *A. vagans* will smother all but the toughest opponents, for *vagans* (roving or wandering) is a name well suited to this extremely hardy and fast-spreading species. It can be used for stabilizing loose crumbling banks and screes, as the tough rhizomes knit together to form a close solid mass. In Victorian times this species was used on some large estates to provide game cover.

DISTINGUISHING FEATURES The dwarf habit of growth, the withering of the leaf-tips and margins during the winter months, and the rampant rootstock. Can be differentiated from *A. humilis* by the leaves, which are much more hairy, and from *A. pumila* by the absence of the ring of hairs at the base of the stem sheaths.

CANES Up to $3\frac{1}{2}$ feet in height by $\frac{1}{2}$ inch in diameter. Bright green at first, maturing to a deep olive green.

NEW SHOOTS Appear from April onwards. Bright green, faintly flushed and striped purple.

CANE SHEATHS Persistent. No hair at the base of the sheath. Pale green at first, fading to a dull straw shade.

NODES Not prominent. Up to 6 inches apart. Bloomed conspicuously on the lower portion with a loose white powder.

BRANCHES Usually borne singly, occasionally in pairs. Fairly long for the length of the cane.

LEAVES Up to 6 inches in length by $\frac{3}{4}$ inch wide. The widths can vary considerably, however, and often reach $1\frac{1}{2}$ inches. Rounded at the base, and narrows to a short sharp point at the tip. Leaf-stalk short and squat. Edged with fine bristles on both margins. Downy on both leaf surfaces, particularly so near the base on the under-surface. Midrib prominent, yellow in colour, and palpable for the whole of its length.

Tessellation clear. Four to eight pairs of secondary veins. Mid-green in colour on the upper surface, dull matt greyish green on the lower. The leaf-tips and margins tend to wither or bleach during the colder winter months, giving the plant a somewhat false appearance of variegation.

PROPAGATION Rhizome cuttings. Division of the clump.

ROOTSTOCK Running. This species is the most invasive bamboo of them all, and if allowed to get out of control can become quite a serious pest. Must be kept to the more unwanted portions of the garden.

Phyllostachys

The obvious recognizable differences between the Phyllostachys and the other types of hardy bamboo, excluding the finer morphological characters of inflorescence, etc., are:

> the canes, which are usually stouter than those of the Arundinaria and show a tendency to zigzag;
>
> the internodes of the canes, branches, and rhizomes, which are channelled or flattened on alternate sides;
>
> the sequence of branch emergence, which is from base to summit;
>
> the branches, which are generally borne in sub-equal pairs with a smaller one between them;
>
> the short-lived non-persistent cane sheaths;
>
> the slow action of the rootstock, which is generally clump forming.

Phyllostachys aurea

Syn: *Bambusa aurea*
    Phyllostachys aureus
    Phyllostachys bambusoides var. *aurea*
    Phyllostachys reticulata var. *aurea*
    Sinarundinaria aurea

Countries of origin: China, Japan
Native names: Gosan-chiku; Hotei-chiku; Kasan-chiku; Taibo-chiku

The habit of growth of this curious bamboo is stiff and erect, but the specific name *aurea* (golden) given it by Carriere is rather misleading, as the canes are usually more green in colour than golden; this is particularly so when the plant is grown in the shade. It is very resistant to frost and drought, and the clumps at Pitt White stood up well to temperatures as low as 11°F during the severe winter of 1962/3.

Some authorities believe that it originated in China, and some say Japan, and it certainly has been in culvitation in Japan, particularly in the Kyushu region, for many centuries. It is known by many names, notably as Hotei-chiku, the 'bamboo of fairyland', and as Taibo-chiku, the 'phoenix bamboo', two delightful descriptive names indeed.

This unusual species has been cultivated in Europe from the middle seventies of the last century, and many plants of diverse clonal ancestry reflect the differences in their parentage by the varying widths and depths of the nodal swellings.

P. aurea is on record as being a species which frequently throws flowering shoots, but the flowering is invariably partial. Some of the flowerings recorded are: England in 1876; France, Belgium, and England in 1904, 1921/2, and in 1935/6.

It is cultivated in the Orient to provide material for walking-sticks, handles for parasols and umbrellas, etc.

P. aurea is believed to be the first Phyllostachys bamboo brought into the United States of America, and introduced originally by a Mr. G. H. Todd of Montgomery, Alabama, in 1822. It has long been known as the 'fishpole' bamboo in that country, owing to the popularity of the canes for fishing rods. Robert A. Young, in his handbook *Bamboo in the United States*, states that this species has withstood temperatures as low as 0°F in certain areas.

DISTINGUISHING FEATURES The curious cylindrical-shaped swellings below the joint, and the close crowding together of the joints towards the base of the cane, give a remarkable appearance, strongly suggestive of something to grip, such as the handle of a weapon, an umbrella, or a walking-stick. It is the only hardy bamboo which has these singular distinguishing features.

CANES Up to 12 feet in height by 1½ inches in diameter. Growth stiff. The base of the cane is thick and it tapers gradually to a slender tip. The internodes are prominently grooved, and the nodes conspicuously swollen. Bright green at first, maturing to a pale creamy yellow. In exposed positions, especially when grown in a site open to the full sun, the cane colouring becomes dull yellow.

NEW SHOOTS Appear from late May onwards. Slightly flattened. Heavily covered with dense layers of coarse sheaths. Olive green to rosy buff, marked with small brown dots. Edible.

CANE SHEATHS Persistent only at the base of the cane, where they overlap closely. They may remain attached to the basal portion for many years. There are two tufts of bristles at the tip. The blade is short,

P. aurea

smooth, and usually reflexed. Rosy buff coloured, with violet streaks, ageing to dull yellow.

NODES Crowded closely at the base. The spacing varies from 1 to 3 inches on the lower extremities, to 6 to 8 inches on the upper parts. Immediately below each joint there is a cylindrical swelling, not unlike an egg-cup in shape, extending downwards for varying lengths, depending upon the diameter of the cane. Smaller versions of these swellings can also be seen on the branch nodes.

BRANCHES Usually borne in pairs, one long and one short at each node. There occasionally is a third branch, but it usually fails to mature and drops off. Stiff and upright in habit. Deeply grooved in the internodes.

LEAVES From 4 to 7 inches in length by $\frac{3}{4}$ inch wide. The base is broadly pointed, and the tip has a short but slender point. Edged with fine bristles on one margin, partially so on the other. Smooth on both surfaces. Midrib fine. Leafstalk fairly long and yellowish green. Tessellation fair. Four to six pairs of secondary veins. Bright light pea green on the upper surface, sea green on the lower.

PROPAGATION Basal cane cuttings. Division of the clump.

ROOTSTOCK Though classed botanically as running, it is, in fact, slow moving and shallow rooted. Makes a good specimen plant.

PHYLLOSTACHYS BAMBUSOIDES (BAMBUSOIDES GROUP)

In the past the term *bambusoides* has been applied to widely dissimilar bamboos, but contemporary botanists and horticulturists now reserve the term for a group of species which includes five easily distinguishable members, and one or two lesser-known variants. *P. quilioi* is regarded by the majority as the type species of the *bambusoides* group. The modern classification is as follows:

Type form: *Phyllostachys quilioi*
Other members of the
    bambusoides group: P. *castillonis*
                  P. *castillonis* var. *inversus*
                    (see under 'Lesser-known hardy groups')
                  P. *marliacea*
                  P. *aurea* and P. *sulphurea* are also considered
                    by some authorities to belong to the
                    group also.

P. castillonis and *P. castillonis* var *inversus* are now considered to have been cultivated variants in the gardens of China. These two variants were introduced into Japan at an early date, brought to France in 1886, and brought to English gardens four years later. The mysterious and elusive bamboo *P. heterocycla* is also believed to belong to this group, and to be another Chinese variant.

Phyllostachys boryana
Syn. *Bambusa boryana*
    Phyllostachys puberula var. *boryana*
    Phyllostachys boryanus
    Phyllostachys niger boryanus
    Phyllostachys nigra cv. 'Bory'
Country of origin: Japan
Native names: Maradake; Unmou-chiku

P. boryana, from Japan, is a member of the *puberula* group, having the same delicately arching canes: but differs in one respect, as the foliage is definitely lanceolate in shape, and not linear, as other members of the group are.

Camus states in his *Les Bambusees* that *P. boryana* flowered in England, France, Germany, and in Switzerland in 1904/5, and although severely damaged by the heavy blossoming, managed to re-establish itself by 1913.

A good specimen plant which shows up well against a dark background. Scarce to come by at one time, but now stocked by most of the leading growers.

DISTINGUISHING FEATURES Similar in growth and habit to *P. henonis,* but has golden-coloured canes. The internodes are marked distinctively with brown, black, or dark purple blotches.

CANES Up to 14 feet in height by $\frac{1}{2}$ to $\frac{3}{4}$ inch in diameter. Slender and arching in habit. Mid-green at first, ripening to a rich yellow splashed with brown, black, or dark purple blotches or spots. The extra internodal colours often run together to form one complete solid colour other than the basic yellow.

NEW SHOOTS Appear from early May onwards. Violet brown. Edible.

CANE SHEATHS Not persistent. Have a kris-shaped blade roughly 1 inch in length. Violet at first, fading to a dull yellow.

NODES Fairly prominent. Up to 10 inches apart. Ringed under the nodes with a prominent white bloom, particularly noticeable when the sheaths fall.

BRANCHES Usually borne in threes at each joint; one long, one short, and one medium length. The smallest branch often fails to reach maturity and is discarded. Branched from base to tip. Slender and tapering. Branchlets numerous and slender.

LEAVES Average 2½ inches in length by ½ inch wide, but may be slightly larger on the younger canes. Base bluntly rounded, and begins to taper 1 inch from the base to end in a very fine tip. Edged with fine bristles on both margins. Midrib fine. Leafstalk long for the size of the leaf. Tessellation minute but clear. Four to seven pairs of secondary veins. Pale yellowish to yellowish green on the upper surface, matt sea green on the lower.

PROPAGATION Division of the clump. Rhizome cuttings.

ROOTSTOCK Running. A fairly slow mover, and makes an excellent spot specimen plant.

Phyllostachys castillonis

Syn. *Bambusa castillonis*
    Phyllostachys bambusoides cv. 'Castillon'
    Phyllostachys castilloni

Country of origin: China

Native names: Hyondake; Kimmei-chiku; Shimadake

P. castillonis is undoubtably the most impressive member of the Phyllostachys, and certainly one of the easiest to recognize. It is difficult to obtain growing portions of the true type, but it is well worth seeking out, as the tall, thick, golden canes with their green-slashed internodes stand out strikingly in any position in the garden.

Originally a native of China, and believed to be a garden variant, it was introduced many centuries ago into Japan. One of its popular names in China is Kimmei-chiku, or the 'golden brilliant bamboo': a beautiful name for a beautiful bamboo. Its stems neither arch outwards with the careless freedom of the other members of the *bambusoides* group nor do they stand stiffly erect like *P. aurea* or *P. mitis*, but bend gently away from the centre, displaying to the full the intricate and delicate tracery of the basic Phyllostachys motif. Besides being beautiful, it is also hardy, and can be seen growing in many parts of the world. In the United States of America it has been recorded as having tolerated temperatures as low as 0°F. I did not hear of one clump being harmed during the severe winter of 1962/3 in the United Kingdom.

It has been cultivated for many years in Europe, and the Curator of

16.
Hedge of
A. simonii

17a, b. Carved bamboo pot belonging to Dr. Mutch, Pitt White

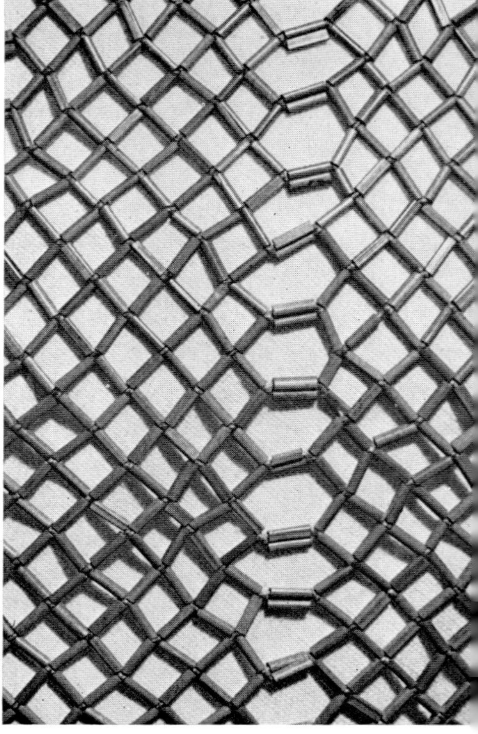

18a, b. A Japanese nobleman's article of court dress made from bamboo.
The enlargement, twice life size, shows the intricate workmanship

the State Nikita Botanic Gardens at Yalta tells me that it flourishes happily there.

It was introduced into France in 1886, and from there to England in 1890. The North American *castillonis* were introduced from France into Florida by a Mr. Henry Nehrling, prior to the bulk introduction of the species from England by the U.S. Department of Agriculture.

P. castillonis flowered at Kew, and at Bicton Gardens in Devon, in 1903/4, and in the 1964/5 seasons in many gardens in the United Kingdom. The latter flowering was prolific, and this, combined with the natural tardiness of new cane production, could well mean the virtual extermination of the species.

In common with its compatriots everywhere else, the clump at Pitt White blossomed heavily, but I was fortunate enough to obtain seed, some of which proved to be viable when sown. The majority of the seedlings, however, turned out to be albinos, and as such lacked the necessary means to manufacture food for growth, and all perished with the exception of six, which were naturally green. By dint of careful nursing I now have six healthy clones. As was generally supposed, *P. castillonis* is almost certainly a variant form of *P. quilioi* (*P. bambusoides*), and the seedlings at Pitt White bear this out. Dr. Stapf of Kew believed that the floral characteristics of this species were akin to the true *P. quilioi*, and I think we can say that he was indeed right.

DISTINGUISHING FEATURES The thick golden-coloured canes, with their vivid bright green deeply etched grooves, which alternate from node to node, make this species instantly recognizable.

CANES Up to 15 feet in height by 1½ inches in diameter. Smooth and glossy. Bright yellow with vivid green grooves. The colour deepens to deep orange yellow when exposed to full sun. Ages to a dull yellow shade.

NEW SHOOTS Appear from late May onwards. Greenish buff coloured, with numerous blotches of brown or brownish purple. Carry a tassel of long brown hairs at the tip. Edible.

CANE SHEATHS Not persistent. Auricles and bristles present on all but the lowest sheaths. Greenish buff at first, ageing to dull brown.

NODES Fairly prominent. The upper ring is dull yellow, and the lower ring is brown. Ten inches to 1½ feet apart.

BRANCHES Usually borne in unequal pairs per node. Upright and arching. Coloured as for the cane.

LEAVES From 3 to 6 inches in length by just over ½ inch wide. Bluntly rounded at the base, and shortly tapered at the tip. One margin edged with fine bristles, the other partially so. Tessellation clear but minute. Six to eight pairs of secondary veins. Bright light green on the upper surface, dull matt green on the lower. Variegation in the leaves is not uncommon in this species, and this usually consists of orange streaks or white pinstripes alternating with the basic green colour.

PROPAGATION Basal cane cuttings. Division of the clump.

ROOTSTOCK Running, but tends to spread slowly unless cultivated in a very warm climate, and even then the yearly production of the new shoots tends to be rather sparse. Makes one of the most outstanding ornamental specimen plants.

Phyllostachys flexuosa
Syn. *Phyllostachys flexuosus*
Country of origin: China

P. flexuosa (bent alternately right and left) has a graceful arching habit, as the name suggests. It was introduced from its native habitat, China, into France by the French Société d'Acclimatation in 1864. It is used in France on a fairly large scale for the manufacture of fishing-rods, and for hedging and screening purposes. It has been found to be very cold resistant in North American and United Kingdom gardens.

Messrs. Riviere noted that this species flowered in 1876, the first sign of the approaching inflorescence being the yellowing of the foliage. When the leaves died the flower panicles began to appear; but Messrs. Riviere also stated that the clumps which flowered did not die, and although greatly weakened, managed to survive and in due course reached their normal proportions.

DISTINGUISHING FEATURES The sheaths of *P. flexuosa* resemble those of *P. viridi glaucescens*, but are shorter and more fragile. They are also less deeply coloured. Can easily be differentiated from *P. viridi glaucescens* by the absence of the side auricles on the cane sheaths, as the latter has either one auricle or a pair per sheath. *P. flexuosa* is decidedly coarser in growth than *P. nigra*, and its canes are always shorter than either *P. nigra* or *P. viridi glaucescens*.

CANES Average 9 to 10 feet in height by $\frac{1}{2}$ inch in diameter. Arching or flexed in habit, and noticeably zigzagged. Bright green at first, becoming golden brown, and often spotted with brown.

NEW SHOOTS Appear from late March onwards. Pale biscuit in colour, spotted with brown, and stained with faint purple streaks. Edible.

CANE SHEATHS Not persistent. Smooth. Bear no side auricles.

Edged with dark brown when young. Blades reddish purple, and horizontally inclined. Greenish beige, edged dark brown and veined purple.

NODES Fairly prominent. Usually 12 to 16 inches apart.

BRANCHES Borne in unequal pairs, one long, one short at each node. Occasionally there may be three branches, but the smallest usually aborts early. Branched from base to summit. Two to three leaves per branchlet.

LEAVES Up to 5 inches in length by ¾ inch wide. Bluntly rounded at the base, and slenderly pointed from half-way to the tip. Leaf-stalk medium long, and fairly thin. Midrib fine, and downy at the base. Fringed with fine bristles on one margin, partially so on the other. Tessellation minute. Six to eight pairs of secondary veins. Dark green in colour on the upper surface, matt greyish green on the underside.

PROPAGATION Basal cane cuttings. Division of the clump.

ROOTSTOCK Running, but fairly static in cooler areas.

Phyllostachys fulva

Syn. *Phyllostachys fulvus*

Country of origin: Japan

P. fulva (reddish or tawny) is a member of the *puberula* group, similar in habit of growth to *P. boryana*, but much more showy. It is relatively rare, and this is unfortunate, as the plant is quite hardy.

A native of Japan originally, it was introduced into England via France in 1898.

DISTINGUISHING FEATURES The canes are distinctive with their tawny colouring, and the pale yellowish-green leaves are also very attractive. The brown line which is often present on the leaf edge is most unusual.

CANES Up to 14 feet in height by ½ inch in diameter. Dull yellow or yellowish green at first, maturing to tawny yellow on one side and brown on the other. Often spattered with brown or black blotches, which may run together to form a solid block of colour.

NEW SHOOTS Appear from late May onwards. Violet brown. Edible.

CANE SHEATHS Not persistent. Violet brown, quickly fading to dull yellow.

NODES Not prominent. Roughly 4 to 6 inches apart.

BRANCHES Usually borne in threes at each node. Branchlets knotted at close intervals on the twigs.

LEAVES Average 3 to 3½ inches in length by ½ inch wide. Rounded at the base, and slenderly pointed at the tip. Edged with fine bristles on both margins. Tessellation good. Midrib fine. Four to six pairs of secondary veins. Mid-green at first, ageing to yellowish green on the upper surface, dull yellowish green on the underside. There is often a brown line down one edge of the leaf.

PROPAGATION Division of the clump. Basal cane cuttings.

ROOTSTOCK Running. Very slow growing, and tends to remain in a fairly circumscribed area. Makes an excellent specimen plant.

Phyllostachys henonis

Syn. *Bambusa henonis*
    Bambusa puberula
    Phyllostachys fauriei
    Phyllostachys nigra cv. 'Henon'
    Phyllostachys nigra var. *henonis*
    Sinarundinaria nigra var. *henonis*
Countries of origin: China, Japan
Native names: Tan-chiku; Makko-chiku; Ha-chiku; Tao-chiku;
    Owodake; Sui-chiku; Suisho-chiku

P. henonis is now regarded by contemporary botanists as the type form of the *puberula* group, and it is widely acclaimed as one of the loveliest of the Phyllostachys. It is noteworthy for the elegance of its slender widely arching canes and the shimmering beauty of its movements when stirred by a breeze. In Japan and China it is known under many names, the most popular being Ha-chiku or 'the light and volatile bamboo'. A beautiful description which matches this graceful bamboo exactly.

When well established it produces regular crops of fresh canes, and not only is it beautiful to look at but the clumps can be planted to make a closely knit screen.

P. henonis is cultivated extensively in most European countries, and in the United States of America and the U.S.S.R. In the United States of America it is known as the 'Henon' bamboo. Its delicate looks are deceptive, for, in fact, it is very hardy, and can tolerate very low extremes in temperature.

Introduced into England in 1890, it began to flower ten years later, and by 1905 many of the original clumps had expired. The survivors gradually grew back to their normal stature, and it is now fairly common in English gardens.

DISTINGUISHING FEATURES The pale rosy-brown, purple-striped new shoots. The very bright green colour of the new canes. The distinct and almost quadrangular tessellation in the leaves. The general delicate appearance of the plant as a whole.

CANES Up to 14 feet in height by $\frac{3}{4}$ to $1\frac{1}{4}$ inches in diameter. Rough in texture at first, becoming smooth later. Have a triple or double groove on alternate sides. No appreciable bloom. Thin walled. Very bright green at first, maturing to a brownish yellow shade.

NEW SHOOTS Appear from June onwards. Pale rosy brown, striped purple. Hairy at first. Edible.

CANE SHEATHS Not persistent. Downy along the edges. Minus auricles. Bright green with reddish brown or tawny splashes at first, quickly fading to dull straw. Blade fairly short.

NODES Not prominent. Five to six inches apart. Rings obvious, the lower being coated with a waxy white bloom. Dark bluish green on the upper part of the node.

BRANCHES Branched from base to tip. Usually three branches per node. The larger branches are solid, with three or four convex sides. Branchlets short jointed, bearing three to four leaves. Purplish in colour.

LEAVES From $3\frac{1}{2}$ to 4 inches in length by $\frac{3}{4}$ inch wide. Midrib fairly prominent. Obtuse at the base, almost linear in shape before tapering off evenly to a very fine point at the tip. Edged with fine bristles on both margins. Leafstalk medium long. Tessellation very distinct and almost quadrangular. Five to seven pairs of secondary veins. Deep brilliant green on the upper surface, light greyish green on the lower.

PROPAGATION Division of the clump. Basal cane cuttings. Rhizome cuttings.

ROOTSTOCK Running. Given warm growing conditions it is fairly free moving, but never invasive. Makes a good specimen plant.

Phyllostachys marliacea

Syn. *Phyllostachys marliaci*
Country of origin: Japan
Native names: Shiwa-chiku; Shibo-chiku

P. marliacea or, as it used to be known, 'Marliac's bamboo', is a fairly hardy but rather rare bamboo. It is known in Japan as the 'wrinkled' bamboo.

Believed to be a variety of *P. quilioi* (*P. bambusoides*), it is very shy in sending up new shoots. These may not appear above the ground until the late autumn, and in colder districts they are invariably cut to the ground by the cold weather. It is therefore a bamboo which should only be planted in those parts which have a late and mild autumn, to allow the new canes enough time to harden before the winter sets in.

I can find no record as to when it was introduced into Europe, nor is there any flowering data.

DISTINGUISHING FEATURES The close crowding of the basal nodes, which makes the base of the cane look wrinkled and gnarled. The smaller branches are distinguishable from those of *P. quilioi* by their slenderness and by their larger quantity of leaves.

CANES Up to 10 feet in height by $\frac{1}{2}$ to $\frac{3}{4}$ inch in diameter. Wrinkled at the base. Dark green, maturing to dull brown.

NEW SHOOTS Appear from late June onwards. Pale pinkish brown, splashed with dark purplish blotches.

CANE SHEATHS Not persistent. Have a conspicuous ring of dark hairs at the base of the blade. Pinkish brown and covered with numerous small brown dots, ageing to a dull straw shade.

NODES Fairly conspicuous. Very closely spaced at the base on mature canes. Often not much further apart than $1\frac{1}{2}$ to 2 inches, much wider on the upper parts of the cane.

BRANCHES Usually borne singly per node, but occasionally in pairs or threes. One branch is always much longer than the others.

LEAVES Up to 5 inches in length by $\frac{3}{4}$ inch wide. Bluntly rounded at the base, and shortly tapered at the tip. The leaf-stalk is relatively short. Edged with fine bristles on both margins. Tessellation fair. Four to five pairs of secondary veins. Bright green on the upper surface, dull greyish green on the lower.

PROPAGATION Basal cane cuttings. Rhizome cuttings. Division of the clump.

ROOTSTOCK Running, but only wanders at the edges and can never be classed as invasive. Makes an excellent specimen plant for an isolated position, and a good ornamental tub plant.

PHYLLOSTACHYS MITIS AND PHYLLOSTACHYS PUBESCENS

These bamboos, both capable of forming groves of considerable height, with stems of timber dimensions in their countries of origin, have often been confused with one another. Camus draws a sharp distinction, reserving the name *pubescens* for the one which is the more massive, but so tender that there is some doubt as to its ability to survive in Northern Europe, and *mitis* for the one in general cultivation in favourable places in France and Belgium.

The same distinction is drawn by the Japanese botanist Honda, and by the American botanist McClure; the latter equates *P. mitis* with *P. viridis* and *P. sulphurea* var. *viridis*, and uses the second and third mentioned names as synonyms.

It is probable that the plants growing out of doors in the United Kingdom are of the hardier species, though *P. pubescens* is known to grow in a few sheltered areas around the southern coasts of Eire and Cornwall.

I propose therefore to deal with these two species under the titles of *P. mitis* (see below) and *P. pubescens* (p. 132) respectively.

Phyllostachys mitis
Syn. *Bambusa mitis*
    Phyllostachys heterocycla
    Phyllostachys sulphurea var. *viridis*
    Phyllostachys viridis
Country of origin: China
Native names: Moso-chiku; Mosodake

P. mitis, a hardy Chinese species, is believed to have been brought from the Orient to Europe *circa* 1864

In the past *P. mitis* has often been confused with *P. pubescens*, and also with *P. sulphurea*, but it has now been established as a separate species by contemporary botanists.

Although *P. mitis* (gentle or without spines) is frost resistant, it is often shy in sending up new shoots during the growing season, and if conditions do not suit it, it may not send any up for many years. Given the right conditions, however, the new shoots will grow with great rapidity. They have been known to grow 6 inches overnight in Cornwall, and in Algiers, where the climate is much warmer, they have been recorded as having grown 20 inches in twenty-four hours.

In warm climates *P. mitis* grows to an immense size, and there are records showing that it has reached 40 feet in height in Italian gardens. McClure states that the canes have reached 40 feet tall by $3\frac{1}{2}$ inches in diameter in certain areas of the south-eastern states of the United States. The canes of *P. mitis* growing in warm climates look like architecturally designed building columns, the cylindrical internodes being held together with thin circular bands of cement, and the shoots are often so far apart that one can walk easily between them.

The new shoots, which are free from bitterness, are prized for their edible qualities, especially the early spring ones. The wood from the mature canes is of good quality, and is used extensively in the Orient for many purposes.

DISTINGUISHING FEATURES Resembles *P. pubescens*, but can be differentiated by the brown-speckled new shoots, the dimpling of the newer cane surfaces, the thick, dull greenish-yellow canes, and the much larger leaves.

CANES Up to 20 feet in height by $1\frac{1}{2}$ inches in diameter. Sub-erect in habit, and slightly curved at the tip. Usually minutely dimpled on the internodal surface when young. Bright green in colour when young, maturing to a dull yellow.

NEW SHOOTS Appear from late May onwards. Yellowish green, spotted with brown or violet brown spots. Edible.

CANE SHEATHS Not persistent. Smooth on the margins. The lowest-borne sheaths are yellow violet in colour. All are blotched or speckled with brown spots. Wither quickly to dull straw.

NODES Not prominent. Just a thin, sharp, straw-coloured ring.

BRANCHES Usually borne in pairs at each node. Branchlets numerous and short jointed. Coloured as the cane.

LEAVES Up to 5 inches in length by up to $\frac{3}{4}$ inch wide. The younger leaves on the new canes may be much larger. Base bluntly rounded, and the tip has a short blunt taper. Leaf-stalk fairly long for the size of the leaf. Edged with fine bristles on one margin, partially so on the other. Midrib fine. Tessellation indistinct to the naked eye. Three to six pairs of secondary veins. Bright pea green on the upper surface, dull greyish green on the lower.

PROPAGATION Basal cane cuttings. Division of the clump.

ROOTSTOCK Though classed as running, it travels only slowly in the cooler areas, and in any case it never throws enough shoots to be classed as invasive. Makes an outstanding specimen plant.

Phyllostachys nigra

Syn. *Bambusa nigra*
    Phyllostachys niger
    Phyllostachys puberula var. *nigra*
    Sinarundinaria nigra
Countries of origin: China, Japan
Native names: Kurodake; Kuro-chiku; Shiro-chiku

P. nigra

In the general form of its growth, *P. nigra* (black), one of the varieties in the *puberula* group, with its long arching canes and its tumbling masses of foliage, is very similar to *P. henonis*.

This species is believed to be the first Phyllostachys bamboo introduced into the United Kingdom. Some authorities give the date as 1827; indeed, it is mentioned in the second edition of Loudon's *Arboretum et Fruticetum*, which was published in 1854, as having been growing in the Horticultural Society's gardens for at least ten or twelve years.

P. nigra is perfectly hardy in almost all temperate areas, and the large clump at Pitt White endured temperatures as low as 19°F for several months without suffering damage. It is cultivated extensively abroad, particularly in the United States of America, and in the Crimea in the U.S.S.R.

There have been many instances of *P. nigra* having been in flower. It did so in England in 1901, and is believed to have done so at Kew nine years previously. In the 1935/6 seasons it flowered in many areas, including Italy, China, the U.S.A., France, and the United Kingdom. In spite of the prolificity of these flowerings, the species has continued to thrive in all these countries. Nevertheless seedlings raised from seed obtained from these flowerings have given rise to a great number of different clonal forms, with canes ranging in colour from dull green to near black. The true *P. nigra* alone possesses canes which are deep rich black.

The rhizomes of *P. nigra* were used, and I have no doubt still are used, in the manufacture of the tough Wang-hai canes, used by walking-stick and umbrella-handle manufacturers.

DISTINGUISHING FEATURES The outstanding feature of this very pretty bamboo is the black canes. There are several partially black-stemmed bamboos, but *P. nigra* is the only one which provides canes which are an even solid black.

CANES Up to 20 feet in height by ¾ to 1½ inches in diameter. Smooth textured. Deeply double grooved in the internodes. Thin walled. Heavily bloomed with a fine white powder under the nodes in the early stages of growth. Olive green in colour during the first season, mottled with black blotches during the second, and maturing to an even solid black.

NEW SHOOTS Appear from May onwards. Pale creamy green flushed with purple and brown. Edible.

CANE SHEATHS Not persistent. Have a few erect bristles at the tip. Hairy during the first part of the season, smooth later. Pale pink, fading to a dull straw shade. The sheath blades are green, and broadly triangular in shape.

NODES Fairly prominent. From 10 to 16 inches apart. Firmly fashioned and delicately outlined. On the mature cane the protruding sharp rim limiting the nodal area is etched in a whitish-grey bloom. The fine white line encircling the lower part of the joint gives added contrast to the glossy black cane. The upper ring of each pair is olive green at first, maturing to a deep purple black.

BRANCHES Usually borne in pairs at each node. Branched from base to tip. The twigs and terminal branches are knotted along their length. Greeny black in colour.

LEAVES Up to 5 inches in length by just over ½ inch wide. The base is bluntly rounded, and the whole almost linear in shape, and tapers steadily to a fine tip. Thin and papery in texture. Leaf-stalk thin and of medium length. Smooth on both surfaces. Edged with fine bristles on one margin, partially so on the other. Midrib fine. Tessellation clear and distinct. Four to six pairs of secondary veins. Dark green in colour on the upper surface, matt greyish green on the lower.

PROPAGATION Basal cane cuttings. Division of the clump.

ROOTSTOCK Running. The mature clump at Pitt White has not been observed to do so. Makes one of the best specimen plants for a select garden spot.

PHYLLOSTACHYS PUBERULA (PUBERULA GROUP)

Contemporary botanists class *P. henonis* as the type form of *puberula*, and with it are grouped several horticultural types commonly differentiated as:

Type form: *Phyllostachys henonis*
Other members
  of the group: *P. boryana*
          P. fulva
          P. nigra
          P. nigra f. *muchisasa* (see under 'Lesser-known hardy
            groups')
          P. punctata

E. G. Camus regards *P. henonis* as the type and the others as variants, and this classification is coming to be generally accepted.

These species are not inconveniently invasive, but form delightful clumps of slenderly delicate arching canes, swaying gracefully under the weight of their luxuriant, fine-leafed foliage.

Phyllostachys pubescens

Syn. *Bambusa edulis*
     Phyllostachys edulis
     Sinarundinaria pubescens
Countries of origin: China, Japan
Native names: Mato-chiku; Kovan-chiku; Rito-chiku; Biotan-chiku;
     Bioji-chiku

Known in Japan as the 'noble' bamboo, *P. pubescens* has the same habit of growth, and the same inability to throw shoots when conditions do not suit it, as that other bamboo with which it has often been confused, namely *P. mitis*. *P. pubescens*, however, is not so hardy as *P. mitis*, and can only be grown in districts where the climate is mild and there is freedom from cold winds. Known also as the 'edible' bamboo, although practically all hardy species of bamboo throw shoots which are edible. Some are sweeter than others, and the shoots of *P. pubescens* are rather acrid when raw, but when cooked are fragrant and delicious. *P. pubescens* and *P. mitis* supply the bulk of the bamboo shoots exported from China and Japan.

According to conditions prevailing in the particular area, the cane heights can vary, and in many districts of China and Japan heights of 75 feet and basal circumferences of 24 inches have been recorded. In the south-eastern coastal areas around the Gulf of Mexico in the United States of America heights of 60 feet have been reported.

The rough hairy surface of the new canes can be plainly felt; these hairs disperse during the following season.

The mature canes are used extensively in the Orient in the manufacture of household utensils, and provide first-class paper pulp for the paper industry.

Named as 'edulis' by Carriere in 1866.

DISTINGUISHING FEATURES The hairy (pubescent) surface of the new canes. This fact provides the main difference between *P. pubescens* and *P. mitis*. The leaves are also smaller, and the branches shorter. The canes are more brightly coloured than those of *P. mitis*.

CANES Up to 15 feet by $1\frac{1}{2}$ inches in diameter in cooler areas. The mature canes tend to bend and twist at the tip. Covered during the first season of growth with a velvety down which makes the surface of the cane quite rough to the touch. Greyish green in colour at first, maturing to a golden yellow. The colouring can become a deep orange if grown in an open sunny position.

NEW SHOOTS Appear from late April/early May onwards. Greenish yellow spotted with red or brown splashes. Edible.

CANE SHEATHS Not persistent. Thick and triangular-shaped, often covering the internode. Hairy. Have a tuft of bristles on each side of the tip. Greenish brown in colour, often splashed with violet spots. Mature quickly to a dull straw shade.

NODES Not prominent. Conspicuously waxy on the lower side, which is white at first, then grey. Resemble those of *P. mitis*, as they are reduced to narrow bands, which on the mature canes scarcely break the contours of the stem.

BRANCHES Normally grown in pairs, or occasionally in threes, but when three appear the smaller branch usually fails to reach maturity and is discarded. Slightly grooved, but often flat faced. Erect in habit with a downward twist at the tip. Branchlets short, progressively diminishing in size towards the tip. Yellowish green in colour.

LEAVES The leaves on mature canes are very small indeed compared to the cane size, usually only 2 to 3 inches in length by $\frac{1}{2}$ inch wide. The leaves on the younger canes can be much larger, but are soon superseded by smaller ones. Bluntly wedge-shaped at the base, and coming to a long slender point at the tip. Edged with fine bristles on both margins. Tessellation very fine, but clear. Midrib fine, and prominent from half-way from the base. Three to five pairs of secondary veins. Light to mid-green in colour on the upper surface, light matt green on the lower.

PROPAGATION Basal cane cuttings. Division of the clump.

ROOTSTOCK Though classed as running, it is practically static in habit in cooler areas.

Phyllostachys punctata

Syn. *Phyllostachys nigra* var. *punctata*
    Phyllostachys puberula
    Phyllostachys punctatus
Country of origin: China

P. punctata was introduced into Europe in the early years of the present century. Once established, this species is very tolerant to low temperatures, but it strongly dislikes drought conditions.

It has flowered all over Europe on several occasions, the most prolific occasions being in 1900 and at Kew in 1908.

There has been much speculation over this species, several authorities believing that it is a clonal seedling of *P. nigra*, but with less colour in its make-up, whilst others say that it is a species in its own right. I leave the correct classification to the botanists, but I can safely say that it does not make as good a specimen plant as *P. nigra* or *P. boryana*, and if given the choice, I would certainly choose one of the latter in preference to *P. punctata*.

DISTINGUISHING FEATURES Can be distinguished from the others in the *puberula* group quite easily by the cane colour (see below). *P. henonis* has bright green canes, and *P. boryana* has canes which age to yellow, as do those of *P. fulva*. The other member of the group, *P. nigra*, of course, has deep black canes.

CANES Up to 12 feet in height by $\frac{1}{2}$ inch in diameter. Muddy green in colour, mottled with irregular black blotches. Very often there are internodes wholly of one colour entirely different from the basic cane colour.

NEW SHOOTS Appear from late May onwards. Olive green, streaked and tipped with faint purple. Edible.

CANE SHEATHS Not persistent. Dull greenish purple, ageing to dull straw.

NODES Upper ring prominent and straw coloured. Average 10 inches apart on the higher parts of the cane.

BRANCHES Normally borne in threes, one short, one medium length, and one long. Branched from base to tip. Branches short jointed, and branchlets numerous.

LEAVES Up to 4 inches in length by $\frac{1}{2}$ inch wide. Leaf-stalk fairly long and thin. Base bluntly wedge-shaped, and tapers sharply from about half-way along the leaf to end in a short sharp tip. Edged with fine bristles on one margin, partially so on the other. Smooth on both surfaces. Tessellation clear. Four to six pairs of secondary veins. Mid-green in colour on the upper surface, duller matt green on the lower.

PROPAGATION Division of the clump. Basal cane cuttings. Rhizome cuttings.

ROOTSTOCK Running. Fairly slow moving, but throws many shoots every season.

Phyllostachys quilioi

Syn. *Bambusa mazeli*
   Bambusa quilioi
   Phyllostachys bambusoides

Countries of origin: China, Japan

Native names: Madake, Yajino (Japan); Nigadake, Karadake, Koku-
   radake (Old Japan); Ku-chiku (China); Deo-bih
   (Assam)

P. quilioi is a very hardy bamboo which will grow and thrive in almost every temperate district. Believed to have been originally a native of China, it has been in cultivation in Japan for many centuries, and has naturalized freely in India and Burma. It is known in Japan as the 'hardy timber bamboo' and as the 'giant timber bamboo'. The specific name *quilioi* was given it by M. Carriere, in honour of the French Admiral Du Quilio, who brought the plant from Japan to France in 1866. It was, and indeed still is, known in some districts of France as Mazel's bamboo, after M. Mazel in whose garden it grew originally in Anduze. It was introduced into the United States of America and the United Kingdom at roughly the same time (*circa* 1890). It is the largest and most commercially valuable bamboo in both China and Japan.

This species has flowered sporadically in many countries over the last sixty years, and although contemporary botanists refer to *P. quilioi* as the *bambusoides* type species, there are in general cultivation many slight varieties of the original type, possibly due to the large number of seedlings raised from the frequent flowerings.

DISTINGUISHING FEATURES Like most other members of the *bambusoides* group, *P. quilioi* forms clumps of widely bending canes, but it is rather stiffer and more sturdy in growth and appearance than the others. In particular, its leaves are conspicuously larger than any others of the Phyllostachys. In general appearance it resembles *P. viridi glaucescens*, but is easily and clearly recognizable by the greater size of its leaves, and by the purple mottling of its stem sheaths, the stem sheaths of *P. viridi glaucescens* being either diffusely stained or streaked with purple.

The ring of long kinked hairs at the base of the blade is also a distinctive *P. quilioi* feature. All members of the *bambusoides* group have these hairs at the base of the sheath blade, but they are most conspicuous on. *P. quilioi*.

CANES Up to 20 feet in height by $1\frac{1}{4}$ inches in diameter. Very thick walled at the base, often solid on the older canes. Smooth textured and highly polished. Deep shining rich green, stained purple at the nodes at first, colouring to a deep yellow olive shade, and maturing to brown.

NEW SHOOTS Appear from late May onwards. Greenish brown, heavily spotted and blotched with brown. Edible.

CANE SHEATHS Not persistent. Have a circle of long purple or greenish-white kinked hairs at the base of the blade. On the upper portion of the cane the sheaths are often covered with fine dark hairs, which lie flat at first, then stand erect before dropping off. Blade short and stubby. Pinkish brown mottled with purple or black blotches.

NODES Fairly prominent, with a thin waxy band on the underside. Average 12 to 18 inches apart. When grown in a sunny exposed site a purple blotch appears on the part of the node facing the sun. Straw coloured.

BRANCHES Usually borne in pairs at each node. One branch is always much larger than the other, the smaller coming from the base of the larger. Occasionally there is a third branch, but this seldom reaches maturity. Three to five leaves per branch.

LEAVES Up to $6\frac{3}{4}$ inches in length by $1\frac{1}{4}$ inches wide. The base is bluntly rounded or wedge-shaped, and the tip is short, sharp, and curves or twists slightly to one side. Midrib fine. Edged with fine bristles on one margin, partially so on the other. The leaf-stalk is long, longer, in fact, than any others of the Phyllostachys. Tessellation clear and well defined. Five to seven pairs of secondary veins. Dark green on the upper surface, sea green on the lower. Some leaves may also be speckled with brown spots.

PROPAGATION Division of the clump. Basal cane cuttings. Rhizome cuttings.

ROOTSTOCK Running. In cooler areas it is rather slow moving, and is never invasive.

Phyllostachys sulphurea

Syn. *Phyllostachys bambusoides* cv. 'Allgold'
    Phyllostachys sulphureus
    Phyllostachys sulphurea var. *holochrysa*
Country of origin: Japan
Native names: Kin-chiku; Ogon-chiku

P. sulphurea is now cultivated extensively in most temperate countries. In England it has shown remarkable resistance to low temperatures.

There has been considerable confusion caused by the misnaming of this species, and it has frequently been sold as *P. mitis* and *P. flexuosa*, and even sometimes as *P. castillonis*. It is true that there are certain points of similarity between these different species, but a look at the 'distinguishing features' shows the essential differences. Mitford believed that *P. sulphurea* belonged somewhere between *P. mitis* and *P. aurea*, but botanists now place the species in the *bambusoides* group. The latest classification is *P. bambusoides* cv. 'Allgold' by McClure in the *Jour. Arn. Arb*, 37, 193, 1956. Correspondence from the Curator of the State Nikita Botanical Gardens at Yalta in the U.S.S.R. states that the clumps growing in the gardens flowered in 1964, and produced some dozens of good seeds.

DISTINGUISHING FEATURES The golden yellow canes with their thin green internodal groove striping and the yellow colour of the branches. The green striping of the grooves is not constant, and may be lacking on many of the canes produced. The only bamboo with which this species may be confused is *P. castillonis*, but *P. sulphurea* lacks the depth of colour that the canes of *P. castillonis* possess, and also has not the same degree of green on the grooves. It may be confused, at first glance, with *P. mitis* also, as this species is similar in habit, but the canes of *P. sulphurea* are much richer in colour, and a good deal thicker.

CANES 30 feet plus in height by up to $1\frac{3}{4}$ inches in diameter. Pale yellow in colour from an early age, maturing to deep yellow, with twin branch grooves which may be dark green in colour. This internodal colouring is not constant, and in some cases may be lacking entirely.

NEW SHOOTS Appear from late May onwards. Brownish yellow, spotted with brown blotches. Edible.

CANE SHEATHS Not persistent. Carry a ring of hairs at the base of the blade. Brownish yellow, spotted with brown, quickly ageing to a dull straw shade.

NODES Not prominent. On mature canes there may only be a thin ring to denote the node. Up to 17 inches apart.

BRANCHES Usually borne in pairs at each node. Short jointed. Yellow in colour from an early age.

LEAVES Average between 3 and 4 inches in length by $\frac{1}{2}$ inch wide. Base bluntly rounded and starts to taper two-thirds from the base to end in a fine tip. Leaf-stalk short. Edged on both margins with fine bristles. Tessellation good. Three to five pairs of secondary veins. Light pea green in colour on the upper surface, matt light green on the underside. The underside is also rough and downy towards the base.

Occasionally there are variegations in the leaf colouring, usually consisting of white or pale yellowish stripes.

PROPAGATION Basal cane cuttings. Division of the clump.

ROOTSTOCK Running. Spreads very slowly, with wide spaces between canes.

Phyllostachys violescens
Syn: *Phyllostachys violascens*
Country of origin: China

P. violescens (becoming violet coloured) is reputed to be allied to *P. viridi glaucescens*, but is not as hardy as that species. It is a bamboo which is seldom seen in English gardens, though several similar species are sold under its label. It was introduced into Europe (France, Germany, and Belgium) in 1869.

The colour pattern evolves in direct contrast to that of *P. nigra*, whose canes proceed from green to black, in that the canes of this species are deep violet when young, gradually changing to greenish yellow.

DISTINGUISHING FEATURES Easily recognized by the deep violet colour of the canes, sheaths, and branches when young.

CANES Up to 14 feet in height by $\frac{3}{4}$ inch in diameter. Bent slightly at the tip. Deep violet in colour at first, ripening to a dull greenish yellow. The colour of the young canes is often so deep that they appear to be black.

NEW SHOOTS Appear from late April onwards. Deep violet in colour.

CANE SHEATHS Not persistent. Have a ring of coarse purple hairs at the tip. Deep violet at first, becoming yellow.

NODES Not prominent. Up to 10 inches apart.

BRANCHES Usually borne in threes of unequal length at each node. Four to five leaves per branch.

LEAVES Average 5 inches in length by $\frac{1}{2}$ inch wide, but can reach up to 7 inches by $\frac{3}{4}$ inch on the younger canes. Bluntly wedge-shaped at the base, and tapered to a fine sharp point at the tip. Edged with fine bristles on both margins. Leaf-stalk short, violet in colour. Tessellation good. Three to six pairs of secondary veins. Striking dark green in colour on the upper surface, dull bluish green on the underside.

PROPAGATION Division of the clump. Rhizome cuttings.

ROOTSTOCK Running. Once established it can spread rapidly, but cannot really be called invasive.

Phyllostachys viridi glaucescens

Syn. *Bambusa viridi glaucescens*
Country of origin: China

P. viridi glaucescens is a graceful and extremely cold-resistant bamboo. In colder areas the canes tend to be rather stiff in habit, but in warmer climates the outer canes are noticeably lax and flexible. The frondage is rather heavy, much heavier than on the other Phyllostachys. This species has flowered in many parts of Europe, and has done so many times during the last fifteen years. I have received information that it bloomed at the State Nikita Botanic Gardens, but few seed were obtained. It is in full bloom in many English gardens at the time of writing.

The American botanist Young writes of this bamboo as having reached the height of 35 feet by 2 inches in diameter, in certain areas of the southern and south-eastern states.

P. viridi glaucescens was introduced into France from China by the French Vice-Admiral Count Cecille in 1846, and shortly after that date it was brought over into England.

DISTINGUISHING FEATURES Recognized primarily by the brilliant blue-green sheen on the underside of the leaves. *P. viridi glaucescens* is also to be differentiated from the other Phyllostachys with similar slender arching canes by the following features: (1) from the *puberula* group, by its larger leaves; (2) from the *bambusoides* group, by its smaller leaves and more slender stems, also by the colour of the cane sheaths, which may be diffusely stained or even lined with purple, whilst those in the *bambusoides* group are mottled with purple spots; (3) from *P. flexuosa*, primarily by the pair of prominent auricles (one of which may be absent) at the apex of the cane sheath, and by the greater length of the canes.

CANES Up to 18 feet in height by $\frac{3}{4}$ inch in cooler areas, much larger in warmer climates. Thin walled. Brilliant green at first, maturing to a dull yellowish-green colour. Often has a violet-coloured prominence on the upper part of the node, very noticeable on the side which is exposed to the full sun.

NEW SHOOTS Appear from early April onwards. Deep green, with long hairs at the tip. Usually spotted with brown or stained violet.

CANE SHEATHS Not persistent. Carries a pair of well-developed auricles, but occasionally can be found with only one. Greenish in colour, spotted with brown and stained violet. Fades to a dull straw shade. The blade is reflexed, and greenish purple.

NODES Very prominent. Usually 6 to 8 inches apart. Tinged violet on the upper prominence.

BRANCHES Usually borne in pairs at each node, but a third may appear, and this one is always much weaker than the other two. Branched from fairly low down the cane. Ultimate branches fairly numerous.

LEAVES From 5 to 6 inches in length by ¾ inch wide. Leaf-stalk long and thin. Shortly wedge-shaped at the base, and long and finely tapered at the tip. Midrib fine. Smooth in texture on the upper surface, downy on the underside, especially near the base. Edged with fine bristles on one margin, partially so on the other. Tessellation very fine. Four to seven pairs of secondary veins. Brilliant green on the upper surface, bluish green on the lower.

PROPAGATION Division of the clump. Basal cane cuttings. Rhizome cuttings.

ROOTSTOCK Running. Wanders freely when grown in mild conditions, but seldom can be classed as invasive. Makes a fairly good specimen plant.

Sasa

The obvious recognizable differences between the Sasa and the other types of hardy bamboo, excluding the finer morphological characters of inflorescence, etc., are:

the dwarf habit of growth;

the leaf size, which is usually greater than on other types of hardy bamboo, and is especially emphasized by the dwarf canes;

the branches, which are usually solitary and only occasionally seen in pairs;

the persistence of the cane sheaths;

the canes, which are usually curved in the lower section;

the marginal leaf withering which distinguishes all the larger-leafed species.

Sasa chrysantha
Syn. *Arundinaria chrysantha*
     Bambusa chrysantha
Country of origin: Japan

S. chrysantha (golden marked) is very hardy, but may be too rampant a grower for the smaller garden, where it will soon take over if allowed to do so. In the larger garden it can, however, be used to cover waste or unsightly ground quickly.

Although named as *chrysantha* by A. B. Freeman-Mitford, it is seldom seen growing as a good variegated clump, for the golden variegations are often absent, and when present are not as good as those on *A. auricoma*, and the latter should always be grown in preference to this species when a golden variegated species is desired.

It is believed that *S. chrysantha* was introduced into Europe in 1892, and no accurate flowering data is available at present.

DISTINGUISHING FEATURES The numerous branchlets. The yellow variegations of the leaves, which is sometimes good, but not as bright nor as conspicuous as the variegations on *A. auricoma*.

CANES Up to 6 feet in height by $\frac{1}{4}$ inch in diameter. Deep olive green.

NEW SHOOTS Appear from late April onwards. Olive green, tipped purple.

CANE SHEATHS Fairly persistent. Very hairy on one side, and have a tuft of hairs at the tip. Dull muddy straw coloured.

NODES Not prominent. Lower rim sharp edged.

BRANCHES Numerous branchlets on the upper nodes. Three to seven leaves per branch.

LEAVES Up to 7 inches in length by 1 inch wide. Rounded at the base, and tapered to a long sharp point at the tip. Smooth on both leaf surfaces. Midrib conspicuous and palpable on the underside. Tessellation clear. Four to six pairs of secondary veins. Bright green on the upper surface, dull matt green on the underside. Streaks of yellow often develop on the leaves; this variegation is unreliable, however, and cannot be trusted to materialize.

PROPAGATION Division of the clump. Rhizome cuttings.

ROOTSTOCK Running. Spreads very rapidly, and can quickly become a nuisance if allowed to run free.

Sasa palmata
Syn. *Arundinaria palmata*
    Bambusa metallica
    Bambusa palmata
Country of origin: Japan
Native names: Chimakizasa; Kumaizasa

S. palmata has the second largest leaves in the hardy Sasa group. As do most of the Sasa, this species tends to suffer from marginal and tip weathering of the leaves during the cold weather. This does the plant little harm, as fresh foliage replaces the affected leaves in the following spring.

Owing to the invasive qualities of this species, it must be classed solely as a bamboo for the larger garden, where it can be allowed to roam freely. I have seen this species used to very good effect when planted on an island, where it was used to provide shelter for the water-fowl.

S. palmata (lobed or divided like a hand) is very hardy, and in its country of origin, Japan, it covers vast tracts in the high mountainous regions. In that country the canes and foliage provide the raw material for the production of hardboard.

Introduced into the United Kingdom in 1889, and into the United States of America in 1925.

At the moment of writing it is in full flower, and has been sporadically flowering since 1963.

DISTINGUISHING FEATURES Differentiated from *S. senanensis* by the smaller leaves, and from *S. tessellata*, which has the largest leaves of the three species.

CANES Up to $7\frac{1}{2}$ feet in height by $\frac{1}{2}$ inch in diameter. Usually curved in growth from the base. Brilliant green in colour, with a waxy efflorescence below the nodes. Matures to dull green.

NEW SHOOTS Appear from April onwards. Pale green, heavily bloomed with a loose white powdery deposit.

CANE SHEATHS Very persistent, often remaining on the cane until ragged. Coarsely striated. Edged with bristles on both margins. Heavily and clearly tessellated. Tongue narrow. Greenish white at first, ageing to a dull drab straw colour.

NODES Not prominent. Average 6 inches apart.

BRANCHES Usually solitary at each node, then one or two secondary branches grow from the primary. Five to nine leaves per branch.

LEAVES Up to 14 inches in length by up to $3\frac{1}{2}$ inches wide. Thick and leathery in texture. Bluntly rounded at the base, and ends in a short point with a fine tail at the tip. Devoid of bristles on the margins. Leaf-stalk thick, short, and wide, yellow in colour. The midrib is prominent and palpable for most of its length, yellow in colour. Tessellation fair. Ten to sixteen pairs of secondary veins. Liable to suffer from marginal and tip withering. Bright pea green on the upper surface, silvery pale greyish green on the lower.

PROPAGATION Rhizome cuttings. Division of the clump.

ROOTSTOCK Extremely rampant in growth, but can be made into a good hedge or screen if the wandering rhizomes can be kept under strict control.

Sasa palmata var. nebulosa

Similar in growth and habit to the species, with the exception of the cane colouring, which is blotched with purple marks instead of the pure brilliant green of the species.

This is a very hardy bamboo, and is popular in gardens all over Europe, and has been sold repeatedly over the years as *S. palmata*. It is slightly less rampant in growth than the species, and it also is in flower at the time of writing.

Sasa senanensis
Syn. *Arundinaria kurilensis* var. *paniculata*
    Bambusa senanensis
    Sasa paniculata
Country of origin: Japan
Native name: Nemagaridake

S. senanensis is a very hardy species from the mountain regions of Hokkaido, Honshu, and Shikaku provinces of Japan. It flowers very frequently and the flower panicles are heavy and coloured a deep purple.

This species is rarely seen in Western gardens and reliable information regarding its growth and habit is not readily available. It is often confused with *S. palmata* and *S. palmata* var. *nebulosa*, and these two bamboos are often sold under the *S. senanensis* label.

Sasa tessellata
Syn. *Arundinaria ragamowski*
    Bambusa ragamowski
    Bambusa tessellata
    Sasamorpha tessellata
Country of origin: China

S. tessellata is a fairly small semi-dwarf but extremely hardy bamboo, and although it has the largest leaves of all the hardy species, it has only small canes to carry them on. The weight of the heavy foliage makes the canes bend outwards, giving the clump a somewhat rounded appearance, reminiscent of a small leafy haystack. In common with the other members of the Sasa group, the leaves wither at the tips and margins during the winter months. It was the first member of the Sasa to be introduced into the United Kingdom, where it arrived from China in 1845.

The name *tessellata* (marked with small irregular squares) is exactly right for this species, as the roughly square-shaped veins are clearly visible to the naked eye. It has often been confused with *A. tessellata* in bamboo literature, and has often been sold as such by mistake. It is, however, entirely different in growth and habit from that species in every possible way, as the canes of *A. tessellata* can grow up to 12 feet plus in height, with leaves of under 6 inches in length, whereas the canes of *S. tessellata* barely top 6 feet, and have leaves of 24 or more inches in length. These differences alone should rule out wrong classification, and a mistake should never be made between these two vastly different species.

DISTINGUISHING FEATURES *S. tessellata* is instantly recognizable by the length of its leaves. It also possesses the largest number of secondary veins in the leaves.

CANES Up to 6 feet in height by $\frac{1}{2}$ inch in diameter. Bloomed with a heavy waxy efflorescence. Bright green in colour.

NEW SHOOTS Appear from May onwards. Usually curved at the base. Pale greenish white in colour.

CANE SHEATHS Very persistent. Pale green at first, fading to dull straw.

NODES Not prominent. Average 6 inches apart.

BRANCHES Usually solitary, but occasionally appear in pairs.

LEAVES Up to 24 inches in length, by up to 4 inches wide. Short abrupt taper at the base, and long and slenderly tapered at the tip. Midrib prominent and palpable for most of its length. Leaf-stalk short, squat, yellow in colour, and often stained with purplish blotches. There is a line of fine hairs on one side of the lower midrib. Tessellation very good. Sixteen to eighteen pairs of secondary veins. Tend to wither at the tips and margins during the winter months. Bright shining green on the upper surface, dull greyish green on the lower.

PROPAGATION Rhizome cuttings. Division of the clump.

ROOTSTOCK Running. Extremely rampant when grown in a warm environment, but fairly slow moving in cooler areas.

Sasa veitchii
Syn. *Arundinaria veitchii*
    Bambusa alba marginata
    Bambusa veitchii
    Phyllostachys bambusoides var. *alba marginata*
    Sasa albo-marginata
Country of origin: Japan
Native name: Kumazasa

This is a hardy dwarf species from the Shikaku, Honshu, and Kyushu provinces of Japan.

The withering of the leaf-tips and margins during the winter months, far from detracting from the beauty of this plant, in fact enhance it, giving it an appearance of variegation. For covering waste ground this species is invaluable, and it will soon smother any weeds in the vicinity.

Introduced into the United Kingdom in 1880.

DISTINGUISHING FEATURES The short canes, the smaller leaves and secondary veins help to distinguish it from other members of the Sasa.

CANES 3 to 4 feet in height by $\frac{1}{4}$ inch in diameter. Thin walled. Heavily bloomed with a white efflorescence. Deep purplish green at first, ripening to a dull purple shade.

NEW SHOOTS Appear from late May onwards. Curved in growth from the base. Covered with a white bloom. Pale green in colour.

CANE SHEATHS Fairly persistent. Have a tuft of coarse hairs at the tip when young. Dull purple, ageing to a dull straw shade.

NODES Inconspicuous. Average 3 to 4 inches apart.

BRANCHES Solitary as a rule, but occasionally appear in pairs. Often the same length as the cane. Four to seven leaves per branch.

LEAVES Up to 10 inches in length by up to $2\frac{1}{4}$ inches wide. Smooth and glossy on the surface. Thick in texture. Midrib fairly prominent, greeny yellow in colour. Bluntly wedge-shaped at the base, and short abrupt tapered at the tip. Leaf-stalk fairly long, squat, and yellow in colour. Both margins fringed with fine bristles. The margins and tips tend to wither during the winter months. Tessellation fine, but clear. Eight to nine pairs of secondary veins. Deep rich green on the upper surface, dull grey green on the lower.

PROPAGATION Rhizome cuttings. Division of the clump.

ROOTSTOCK Running. Moderately invasive, soon forms a dense carpet.

S. veitchii

Sasa veitchii var. nana
Syn. *Sasa veitchii* f. *minor*
Country of origin: Japan
Native name: Yakibazasa

Similar in growth and habit to the species, but the leaves are slightly
smaller and more sharply pointed. The withering or bleaching of the
leaf-tips and margins is also more conspicuous. It is not as aggressive
as the species, and is more easily kept under control.

Grown extensively in France and in the United Kingdom, and should
always be grown in preference to *S. veitchii*.

Bambusa

Bambusa quadrangularis
Syn. *Bambusa angulata*
    Chimonobambusa quadrangularis
    Tetragonacalamus quadrangularis
Country of origin: China
Native names: Shikakudake; Shiho-chiku; Toocu-chu

B. quadrangularis (four cornered), or the 'block' bamboo, as it is often
known, is reputed to be rather tender. This may be so in the very cold
districts, but the clump at Pitt White stood up to 19 degrees of frost for
many months during the big freeze of 1962/3, and suffered only minor
leaf scorch. Even when the weather is so severe as to cut down the
canes, the rhizomes are seldom damaged, and new growths will emerge
in the following spring.

It is a native of China, and has been naturalized in Japan for many
centuries. It also grows freely on Formosa. There is a well-known grove
of this species, some 30 feet in height, in the grounds of a temple at
Osaka. In the cooler climate of western Europe it never reaches these
proportions, nevertheless it is a most attractive bamboo, and suits all
types of garden.

Along with the unusual angular shape of the canes, this bamboo has
another singular feature, in that the branches can be snapped off easily
by hand, unlike most bamboos, of which the branches can only be
severed by secateurs.

The mature canes, though light, are extremely thick walled, and the
wood is tough and close grained; they therefore make admirable plant
supports.

Not only does B. *quadrangularis* make an attractive garden plant, it also makes an excellent tub specimen.

DISTINGUISHING FEATURES The squareness of the mature canes and the rings of embryo buds on the stems at the nodes. Also the wideness of the nodes in contrast to the slender internodal sections.

CANES Up to 10 feet in height by $\frac{3}{4}$ inch in diameter. Obtusely quadrate in cross-section, especially on the older canes, and more pronouncedly around the basal area. Dark green in colour, maturing to brownish green. Occasionally has dark purple blotches.

NEW SHOOTS Appear from late May onwards. Pale straw coloured, heavily overlaid with purple and black blotches. Edible.

CANE SHEATHS Not persistent. Fall very early in the first season. Tessellated. Short. Dull straw coloured, stained with purple splashes on the interior surface.

NODES Very conspicuous. Swollen and sharply projecting, especially on the lower part of the cane, where the space between the paired nodal rings is studded with embryo buds and dormant rootlets. 5 to 6 inches apart.

BRANCHES Numerous. Borne in clusters of three to five at each node, except the lowest ones. Up to 3 feet in length. Two to four leaves per branchlet.

LEAVES Up to 9 inches in length, by up to $1\frac{1}{4}$ inches wide. Sharply wedge-shaped at the base, bending gracefully to a slender taper, and ending in a long tongue at the tip. Leaf-stalk fairly long and thin. Midrib fine, but fairly prominent, and palpable for most of its length. Edged with fine bristles on both margins. Tessellation fairly prominent. Five to seven pairs of secondary veins. Deep olive green in colour on the upper surface, smooth matt green on the lower.

PROPAGATION Rhizome cuttings. Basal cane cuttings. Division of the clump.

ROOTSTOCK Running. Can be very active when grown under favourable conditions. A long runner, but bears only a few canes annually. Canes well spaced. May wander, but unlikely to be classed as invasive.

Chusquea

Chusquea culeou
Syn. *Chusquea breviglumis*
Country of origin: Chile

A. B. Freeman-Mitford listed sixteen species of Chusquea which are

Chusquea culeou

native to South America, Mexico, and the West Indies. Camus describes sixty-nine species, but of these *C. culeou*, which comes from Chile, is at the present time the only one which can be truly classed as hardy.

This species is a fast grower when given the right conditions; shoots have frequently grown 6 inches overnight in the garden at Pitt White. It is also very hardy, and able to stand considerable periods of low temperature. Another singularity is that it is able to withstand fairly long dry spells. The clump in the Royal Botanic Gardens in Edinburgh is considered to be the hardiest in their collection of hardy bamboos. It grows freely in many areas of Europe, and there are many well-established stands in England, notably at Hidecote Manor, at the R.H.S. gardens at Wisley, and at Pitt White.

C. culeou is believed to have been discovered in Chile by the botanist Gay, who named it, and brought it to England in 1890.

There is no record of this species having been in flower outside its native habitat.

DISTINGUISHING FEATURES The clustered branch formations and the stiffly standing bottle-brush appearance. The white sheaths, which in the first season produce a chequerboard effect on the bright green, solid young canes, as they cover either side of the cane alternately. Unlike most other hardy bamboos, the canes of this species are solid throughout and not hollow. This is a bamboo which is outstanding in appearance, and cannot ever be mistaken for any other hardy type.

CANES Up to 18 feet in height by $1\frac{1}{2}$ inches in diameter. Solid. Sturdy in appearance, and tapering to a gradual point at the tip. Branchless and leafless during the first season. Deep olive green in colour during the early years, maturing to a dull yellowish-brown shade.

NEW SHOOTS Appear from late April onwards. Heavily striated. Thick and succulent. Purplish green in colour, heavily overlaid with mauve shadings. Coated with a bluish white bloom. Edible.

CANE SHEATHS Persistent for the first season. Long, often overlap the internodes. Mauve at first, quickly fading to a creamy white.

NODES Fairly prominent. From 4 to 6 inches apart. Have a deep ring of greyish-white bloom on the underside of the node, which quickly disperses during the second season.

BRANCHES Short, usually only 18 inches in length, but may reach $2\frac{1}{2}$ feet. Very slender. Multiple at every node, and extending two-thirds of the way around the cane, giving the thick bottle-brush effect which is characteristic of the species.

LEAVES Up to $3\frac{1}{2}$ inches in length by $\frac{1}{4}$ inch wide. Slightly wedge-shaped at the base, and long and slenderly pointed at the tip. Leaf-stalk very short and thin. Edged with fine bristles on both margins. Midrib fine, but palpable. Tessellation very fine, but clear. Two to six pairs of secondary veins. Deep mid-green on the upper surface, dull matt green on the underside.

PROPAGATION Basal cane cuttings. Division of the clump.

ROOTSTOCK Caespitose. Stays in a close circumscribed clump. Makes an ideal isolated specimen clump.

Shibataea

Shibataea kumasasa
Syn. *Bambusa kumasaca*
  Bambusa viminalis
  Phyllostachys kuma-saca
  Phyllostachys kumasasa
  Phyllostachys ruscifolia
  Sasa ruscifolia
Country of origin: Japan
Native names: Bundodake; Bungozasa; Gomaizasa; Kagurasasa;
    Lyozasa; Okamesaza

The general squatness of the growth of this species makes the Shibataea group quite distinct from the Phyllostachys, to which it used to belong. It was named in honour of the eminent Japanese botanist K. Shibata, who has himself named many species of bamboo. An attractive hardy dwarf bamboo, *Sh. kumasasa* is widely cultivated in western Japan, and it was introduced into the United Kingdom in 1861. It is widely grown in many areas of Europe, and in the United States of America. It requires plenty of moisture, but can resist cold winds and severe frosts.

This species is one of the first to send up its fresh shoots in the spring.

DISTINGUISHING FEATURES The squat broad leaves, and the very dwarf compact habit of growth, are features that separate this species from the other dwarf hardy species.

CANES From 2 to 2½ feet in height by ¼ inch in diameter. Almost solid. Smooth in texture. Very zigzagged in habit of growth. Roughly triangular or slightly elliptical in shape. Pale green at first, maturing to a dull brown colour.

NEW SHOOTS Appear from late March onwards. Pale green, tipped purple.

CANE SHEATHS Fairly persistent. Fringed with fine hairs. Purple in colour, but fading quickly towards the tip to a dull straw shade.

NODES Prominent for the size of the cane. Up to 3 inches apart. Internodes prominently grooved. Dark green in colour.

BRANCHES Usually borne in pairs, but occasionally in threes. Short. One leaf, rarely two, at the apex.

LEAVES Up to 4 inches in length by 1 inch wide, beginning to taper two-thirds of their length from the base to end in a short sharp tip, and broadly pointed at the base. Midrib fine. Leaf-stalk long for the size of the leaf. Edged with fine bristles on both margins. Tessellation conspicuous and very clear. Five to eight pairs of secondary veins. Upper surface dark green in colour at first, but quickly fading to a dull yellowish-green shade; lower surface dull greyish green.

PROPAGATION Division of the clump. Rhizome cuttings.

ROOTSTOCK Running type, but usually forms fairly compact clumps, and seldom becomes invasive.

Lesser-known Hardy Groups

LIST OF HARDY BUT NOT ESTABLISHED BAMBOOS

Arundinaria amabilis
      aristatus
      chino var. argenteostriata
      communis
      fastuosa var. kagamiana
      fastuosa var. yashadake
      intermedia
      khasiana
      kinsiana
      linearis
      marmorea cv. 'Variegata'
      niitakayamensis var. microcarpa
      pantlingii
      simonii var. variegata
      tecta var. decidua
      vaginata
      variegata
      virens

Phyllostachys castillonis var. inversus
        heterocycla cv. 'Kikku-chiku'
        nigra f. muchisasa
        viridi glaucescens var. linculi

Sasa borealis
    cernua
    kurilensis
    nipponica
    owatarii
    ramosa
    uchidae

Bambusa multiplex (typical form)
        multiplex cv. 'Alphonse Karr'
        multiplex cv. 'Fernleaf'
        multiplex var. riviereorum
        multiplex cv. 'Silverstripe'
        multiplex cv. 'Silverstripe Fernleaf'
        multiplex cv. 'Stripestem Fernleaf'
        multiplex cv. 'Willowy'
        tootsik
        vulgaris

Arundinaria amabilis
Country of origin: China
Native name: Commonly known as the 'Tonkin' bamboo

A. amabilis is moderately cold resistant and reports show that it has withstood several degrees of frost in southern parts of the country. The canes, which are tough and pliable, are used extensively in the manufacture of split-cane fishing-rods, and form the bulk of garden canes imported from the Orient.

CANES 15 to 20 feet in height by up to 1 inch in diameter. Mid-green in colour. Thick walled and tough.

BRANCHES Normally one large branch is prominent with one or two laterals emerging from its base.

LEAVES Very variable in size, ranging from 4 to 14 inches in length by $\frac{1}{2}$ to $1\frac{1}{2}$ inches wide. Smooth and glistening on the upper surface, slightly rough on the underside. Bristled on both margins. Bright green on the upper surface, dull matt green on the lower.

Arundinaria aristata

Syn. *Thamnocalamus aristatus*
Country of origin: India

A. aristata (having an awn or beard) is an Indian bamboo from the north-western Himalayas, where it grows at altitudes of between 9,000 and 11,000 feet above sea-level, in the states of Sikkim and Bhutan. In its native habitat it flowers and dies down every year, but in milder climates where the winters are less severe it will retain its canes and lose the annual habit of flowering, developing after a few seasons into a compact clump. When used as a pot plant the canes are dwarfed and are stiffer in growth.

There are numerous reports of this species having been in flower in Western gardens. It flowered in 1889 and 1901 in Gamble's old garden at Liss in Hampshire, and in the Royal Botanic gardens at Kew in 1950.

A. aristata was collected originally from the Himalayas by Kurz in 1868, and by C. B. Clarke in 1869.

DISTINGUISHING FEATURES The brownish-green colouring of the canes. The dull red colour of the branches and branchlets, and the curious twist each leaf has at its tip.

CANES 16 feet plus in height by $\frac{1}{2}$ inch in diameter. Brownish green at first, maturing to a dull yellow shade.

NEW SHOOTS Appear from late May onwards.

CANE SHEATHS Not persistent. From 8 to 12 inches apart.

BRANCHES Numerous branches and branchlets at every node. Dull red.

LEAVES Up to 5 inches in length by $\frac{1}{2}$ inch wide. Bluntly rounded at the base and tapering to a fine twisted point at the tip. Leaf-stalk tiny. Midrib fairly prominent. Edged with fine bristles on both margins. Tessellation fair. Four to five pairs of secondary veins. Bright green in colour on the upper surface, dull greyish green on the lower.

PROPAGATION Division of the clump. Basal cane cuttings.

ROOTSTOCK Caespitose. Makes an excellent specimen plant and a first-rate tub subject.

Arundinaria chino var. argenteostriata

Syn. *Arundinaria argenteostriata*
    Bambusa argenteostriata
    Nipponocalamus argenteostriatus
Country of origin: Japan
Native name: Okinadake

A variety which has no great attraction as a garden plant. Similar to *A. chino*, but has variegated foliage.

Arundinaria communis
Syn. *Arundinaria argenteostriata* var. *communis*
      Pleioblastus communis
Country of origin: Japan
Native name: Gokidake

CANES Up to 9 feet in height by ½ to ¾ inch in diameter. Deep greenish purple.

LEAVES Lanceolate. Vary from three to fifteen per branchlet. Mid-green.

GENERAL INFORMATION Native to the Honshu, Shikaku, and Kyushu districts of Japan.

Arundinaria fastuosa var. kagamiana
Syn. *Semiarundinaria kagamiana*
Country of origin: Japan
Native name: Rikuchudake

GENERAL INFORMATION Differs from the type in floral characteristics, and is also less densely pubescent. Less vigorous in growth. Cultivated in Honshu (Rikuchu).

Arundinaria fastuosa var. yashadake
Syn. *Semiarundinaria fastuosa* var. *yashadake*
      Semiarundinaria yashadake
Country of origin: Japan
Native name: Yashadake

GENERAL INFORMATION Differs from the type by the canes being more slender, the branchlets sparser, the sheaths on the nodes on the lower part of the cane being short and hairy and lack auricles. The leaves are three to five per branchlet, short tapered at the tip. The main feature which differentiates it from the type is the length of the prophylla, which are much longer than those of the type. Originated in the Honshu, Shikaku, and Kyushu districts of Japan.

Arundinaria intermedia
Syn. *Sinobambusa intermedia*
Country of origin: India

This species is semi-tender and should only be grown in areas where there is freedom from severe frosts and cold winds. It comes from the eastern Himalayan areas of Sikkim, where it grows at heights of between 7,000 and 8,000 feet above sea-level.

CANES Up to 12 feet in height by ½ inch in diameter. Mid-green.
NODES Not prominent.
BRANCHES Numerous at every node, giving the plant a tufted appearance.
LEAVES Up to 8 inches in length by 1 inch wide. Hairy on the underside near the base of the midrib. Seven to eight pairs of secondary veins. Bright green on the upper surface, dull greyish green on the lower.
PROPAGATION Basal cane cuttings. Division of the clump.
ROOTSTOCK Caespitose. Makes a good specimen plant if grown in a warm environment. A good tub subject for the greenhouse.

Arundinaria khasiana
Country of origin: India

This unique species is believed to originate from Assam, and like most of the higher-altitude bamboos it does not like the cold winds, and should be grown in climates which have above-the-average temperatures. It is somewhat akin to *A. falcata*, but its habit of growth is much stiffer and stronger than the latter.

DISTINGUISHING FEATURES The deep green, almost black cane colouring. The absence of visible tessellation in the leaves.
CANES From 8 to 12 feet in height by ½ inch in diameter. Deep green, almost black on the older canes.
CANE SHEATHS Not persistent.
NODES Fairly prominent. Roughly 1 foot between nodes.
BRANCHES Numerous, especially on the upper joints.
LEAVES Up to 4 inches in length by ½ inch wide. Edged with fine bristles on one margin, partially so on the other. Slightly downy on the underside. No visible tessellation. Three to four pairs of secondary veins. Pale pea green on the upper surface, dull matt green on the lower.
PROPAGATION Basal cane cuttings. Division of the clump.
ROOTSTOCK Caespitose. Makes a good specimen plant.

Arundinaria kinsiana
Syn: *Pleioblastus kinsianus*
Country of origin: Japan
Native name: Fushiodake-shino

CANES From 6 to 9 feet in height by ½ inch in diameter.
NODES Not conspicuous. Coated with fine bristles.
LEAVES Up to 7 inches in length by ¾ inch wide. Bluntly rounded at the base, long tapered at the tip. Three to eight leaves per branchlet. Mid-green.

Arundinaria linearis
Syn. *Pleioblastus linearis*
Country of origin: Japan
Native names: Gyoyo-chiku; Ryuku-chiku

This species is hardy, but is comparatively rare in Western gardens. It is spontaneous in the Ryuku Islands, and cultivated extensively in the Kyushu district of Japan.
CANES Up to 15 feet in height by ½ inch in diameter. Smooth. Green.
BRANCHES From one to five per node. Densely ramified.
LEAVES From six to ten per branchlet. Dark green.

Arundinaria marmorea cv. 'Variegata'
Syn. *Chimonobambusa marmorea* f. *variegata*
     Chimonobambusa marmorea var. *variegata*
Country of origin: Japan
Native name: Chigo-kan-chiku

Similar in growth and habit to the type, with the exception of the leaves, which have white variegations.

Arundinaria niitakayamensis var. microcarpa
Syn. *Indocalamus niitakayamensis* var. *microcarpa*
Country of origin: Formosa

Nearly identical morphologically in habit and growth to the type, but is believed to have smaller flowers.

Arundinaria pantlingii
Country of origin: India

Comparatively little is known or has been written about this species, but it is known that *A. pantlingii* comes from the Himalayan regions, where it grows at altitudes of between 11,000 and 12,000 feet above sea-level in the states of Sikkim and Bhutan. It is believed to have been cultivated originally in England in a garden at Liss in Hampshire in 1919, its owner having imported the stock from a garden at Darjeeling. The last recorded flowering of this species was in 1895, but it is not believed to have done so as yet in the United Kingdom.

DISTINGUISHING FEATURES The ring of tattered sheath remains which adheres to the nodal ring. The long twisted point at the leaf-tips.

CANES Up to 12 feet in height by ½ inch in diameter. Very thin walled. Dull olive green in colour.

NEW SHOOTS Appear in the latter part of April.

CANE SHEATHS Very persistent. The basal portion will remain attached to the cane long after the main part has withered away. Glossy and striated. Have a long fringe of hairs along each edge, and a pair of prominent auricles with six to ten curved hairs attached. The blade is prominent and downy on both surfaces.

NODES Not prominent. From 6 to 7 inches apart.

BRANCHES Numerous at every joint.

LEAVES Up to 6 inches in length by just over ½ inch wide. Bluntly rounded at the base, and narrows to a long tapering tip which ends in a twisted point. Smooth on both surfaces. Edged with fine bristles on both margins. Tessellation very fine. Four to five pairs of secondary veins. Dull green in colour on the upper surface, matt dull green on the lower.

PROPAGATION Rhizome cuttings. Division of the clump.

ROOTSTOCK Running. Although classed as such, it is usually rather slow in becoming established.

Arundinaria simonii var. variegata

Syn. *Arundinaria simonii* cv. 'Silverstripe'
    Bambusa albo-striata

Countries of origin: China; Japan

This species is identical in habit to the type, the only distinction being the leaf colouring. This is deep green on the larger leaves, but on the smaller leaves there is frequently white striping. The striping is often extensive, but is almost always confined to the smaller leaves.

It in no way compares with *A. fortunei* in the white variegations, and the latter should always be planted in preference.

Although quite hardy, it is not as tall as the type, and it cannot be recommended as a specimen plant.

Along with *A. simonii* it flowered at the State Nikita Botanical gardens at Yalta in the U.S.S.R., but did not set seed.

Arundinaria tecta var. decidua
Country of origin: The United States of America

This bamboo is rather uncommon outside its native habitat, and is similar in growth to the type, but differs in that the foliage ripens to yellow in the autumn, then falls. It is slightly smaller than *A. tecta*. It was discovered growing in the Biltmore and Ashville areas of western North Carolina by C. D. Beadle in the early part of this century.

Arundinaria vaginata
Syn. *Nipponocalamus vaginatus*
     Pleioblastus vaginatus
Country of origin: Japan
Native name: Hakonedake

CANES Up to 12 feet in height by $\frac{1}{2}$ inch in diameter. Dull green.
BRANCHES Usually in threes at each node.
LEAVES Up to 8 inches in length by $\frac{1}{2}$ inch wide. Bluntly rounded at the base, and sharply pointed at the tip. Mid-green in colour.

Arundinaria variegata
Syn. *Sasa variegata*
Country of origin: Japan

There has been much speculation as to the true identity of this bamboo, but it is widely grown in the U.S.A., where it is known as the 'dwarf whitestripe' bamboo.

CANES Up to 3 feet in height by $\frac{1}{4}$ inch in diameter. Green.
BRANCHES Usually borne singly, but occasionally appear in pairs. Often have a secondary branch at the base of the primary.
LEAVES Up to 6 inches in length by 1 inch wide. Coated with fine hairs on the underside. Base rounded, tip short, sharply pointed. Usually from five to ten leaves per branch, gathered together at the tip. Deep green, striped white or cream.

Phyllostachys castillonis var. inversus

The growing habit of this bamboo is identical to the type, but the cane colour is reversed, i.e. the grooves are yellow and the cane colour is green. It is a hardy bamboo, but unfortunately now rather rare. Believed to be of Chinese origin and a garden variant of *P. castillonis*, which in its turn is reputed to be a variant of *P. quilioi* (*P. bambusoides*).

Arundinaria virens
Syn. *Nipponocalamus virens*
    Pleioblastus virens
Country of origin: Japan
Native name: Ao-nezasa

CANES Up to 12 feet in height by ½ inch in diameter. Mid-green in colour.

LEAVES Roughly 8 inches in length by up to 1 inch wide. Rounded at the base, and long pointed at the tip. Bright green.

Phyllostachys heterocycla cv. 'Kikku-chiku'
Country of origin: China
Native names: Lohan-chu; Kikku-chiku

The so-called 'tortoise-shell' bamboo which is described in Freeman-Mitford's *The Bamboo Garden* is believed by some authorities to be a garden variant of *P. aurea*, and Young states that the nodes of *P. aurea* are often found to be obliquely inclined, forming attractive designs simulating tortoise-shell patterning. Camus, on the other hand, held that it was botanically related to *P. mitis*, but more stable in its characters.

The monstrous obliquely inclined nodes are not a permanent feature, and occur only on isolated canes in a clump. It is a botanical curiosity seldom seen in Western gardens, and has not been seen in England since Freeman-Mitford's time.

Phyllostachys nigra f. muchisasa

This bamboo is believed to be a seedling of *P. nigra*, and though similar in habit of growth, it lacks the dense black cane colour. It has often been sold as the type, and is quite common in many areas of Europe.

Phyllostachys viridi glaucescens var. linculi

Practically identical to the type, and differs only in minute morphological characters.

Sasa borealis

Syn. *Arundinaria purpurascens*
    Bambusa borealis
    Pseudosasa spiculosa
    Sasamorpha purpurascens
    Sasa purpurascens
    Sasa purpurascens var. *borealis*
    Sasa spiculosa
Country of origin: Japan
Native names: Jidake; Suzudake

S. borealis, a hardy dwarf bamboo, comes from the Hokkaido, Honshu, Shikaku, and Kyushu districts of Japan.

CANES Up to 3½ feet in height by ¼ inch in diameter. Dull purplish green in colour.

NODES Not prominent.

LEAVES Up to 12 inches in length by 1½ to 2 inches wide. Short tapered at the base, and short but sharply tapered at the tip. Lustrous mid-green in colour on the upper surface, dull matt green on the lower.

Sasa cernua

Syn. *Sasa kurilensis* var. *cernua*
Country of origin: Japan
Native name: Okumyamazasa

A dwarf hardy bamboo from the Hokkaido and Honshu districts of Japan. The thick heavy leaves tend to make the canes droop, thus the name *cernua*, or drooping.

CANES Up to 3½ feet in height by ½ inch in diameter. Mid-green in colour.

NODES Fairly prominent.

LEAVES Up to 7 inches in length by 1½ to 2½ inches wide. Thick and leathery. Hairy on the underside. Broadly rounded at the base, and bluntly tapered at the tip. Drooping habit.

Sasa kurilensis
Syn. *Arundinaria kurilensis*
    Bambusa kurilensis
    Pseudosasa kurilensis
Countries of origin: The Kurile Islands; Sakalin; Korea
Native name: Chishimazasa

A hardy, high mountain dwarf bamboo, found originally in the Kuriles, and now naturalized in the Hokkaido and Honshu districts of Japan.
CANES Up to 6 feet in height in warm areas, but smaller in cooler districts, usually $\frac{1}{2}$ inch in diameter. Dull green in colour.
BRANCHES Very sparse.
LEAVES Up to 8 inches in length by $\frac{1}{2}$ inch wide. Lustrous mid-green in colour.

Sasa nipponica
Syn. *Bambusa nipponica*
Country of origin: Japan
Native name: Mizakozasa

A hardy dwarf species from the Honshu, Kyushu, and Shikoku districts of Japan.
CANES Up to 4 feet in height by $\frac{1}{2}$ inch in diameter. Green.
BRANCHES Usually solitary. Sparsely branched at the base.
LEAVES Linear oblong. Up to 10 inches in length by $1\frac{1}{4}$ inches wide. Bluntly pointed at the base, and sharply tapered at the tip. Auricles rounded. Mid-green in colour.

Sasa owatarii
Syn. *Arundinaria owatarii*
    Pseudosasa owatarii
    Yadakeya owatarii
Country of origin: Japan
Native name: Yakushimadake

A hardy dwarf bamboo from the high mountain areas of the Kyushu district of Japan.
CANES Up to 4 feet in height by $\frac{1}{2}$ to $\frac{3}{4}$ inch in diameter. Deep green in colour.
BRANCHES Branched on the upper part of the cane only.

LEAVES Up to 6 inches in length by ½ inch wide. Bluntly rounded at the base, and long tailed at the tip. Oral setae lacking. Tessellation prominent. Yellowish green in colour.

Sasa ramosa
Syn. *Arundinaria ramosa*
Bambusa ramosa
Sasaella ramosa
Country of origin: Japan
Native name: Azumazasa

A medium-dwarf hardy bamboo from the Honshu district of Japan.
CANES Up to 6 feet in height by ½ inch in diameter. Curved in growth at the base. Purplish in colour.
BRANCHES Solitary on the upper parts. Three to five leaves per branch.
LEAVES Up to 6 inches in length by ¾ inch wide. Rounded at the base, and bluntly tapered at the tip. Thick and leathery. Mid-green in colour.

Sasa uchidae
Syn. *Pseudosasa uchidae*
Sasa kurilensis var. *uchidae*
Country of origin: Japan
Native name: Nagaba-nemagaridake

A hardy dwarf bamboo from the Hokkaido and Honshu districts of Japan. Somewhat similar in growth and habit to *S. kurilensis*, but the leaves are smaller.

Bambusa multiplex (typical form)
Syn. *Bambusa argentea*
Bambusa nana
Bambusa nana var. *normalis*
Bambusa nana var. *argentea*
Leleba multiplex
Country of origin: China
Native name: Horai-chiku

B. multiplex (manifold) originally came from the Chinese province of Kwangtung. It is fairly hardy, and though rare in Europe it is one of the commonest species under general cultivation in the U.S.A., where it is known as the 'hedge' or the 'Oriental hedge bamboo'. In California it is reported as having stood up to temperatures as low as 17°F.

It is very variable in cultivation, and there are at least eight recognized forms of this bamboo in existence.

DISTINGUISHING FEATURES The thin-walled, plain green-coloured canes. The silvery sheen on the underside of the leaf, and the ear-shaped sheaths which are typical of the Bambusa.

CANES Very variable in size, ranging from 10 feet to 35 feet in warm areas, by $\frac{1}{2}$ to $1\frac{1}{2}$ inches in diameter. Thin walled. Arched in habit. Dull plain green, maturing to dull yellow.

NEW SHOOTS Appear in the late spring.

CANE SHEATHS Persistent. Slightly hairy. The blade drops off when dry. Dull green fading to yellow.

NODES Rather prominent and thickened.

BRANCHES Absent during the first season, multiple thereafter. One branch is large with two roughly equal medium-sized ones, the remainder being small to tiny.

LEAVES Up to 4 inches in length by $\frac{1}{4}$ to $\frac{1}{2}$ inch wide. Bluntly rounded at the base, short tapered at the tip. Midrib faint. Tessellation not visible. Three to six pairs of secondary veins. Plain green in colour, silvery green on the underside.

ROOTSTOCK Caespitose. Creeps along very slowly. Makes a good hedging and screening plant.

PROPAGATION Division of the clump. Basal cane cuttings.

Bambusa multiplex cv. 'Alphonse Karr'

Syn. *Bambusa alphonse karri*

    Bambusa nana var. *alphonse karri*

    Leleba multiplex f. *alphonse karri*

Country of origin: China

Native name: Sou-chiku

The canes of this variety can top 30 feet in height, and are coloured bright yellow and striped with vivid green. The cane sheaths are also colourful, being dark brownish green, striped with white and yellow streaks. The streaks remain after the other colours have faded. Can tolerate fairly low temperatures. Named after Jean Baptiste Alphonse Karr, a noted French horticulturist of the nineteenth century.

Bambusa multiplex cv. 'Fernleaf'
Syn. *Bambusa nana* var. *disticha*
    Bambusa nana var. *gracilloina*
    Leleba floribunda
Country of origin: China
Native name: Ho-o-chiku

Canes can be as tall as 20 feet and are coloured as the typical form. Leaves usually small, borne in pairs, plain green in colour. Up to twenty per twig.

Known as the 'fernleaf' hedge bamboo in the U.S.A., where it has been recorded as having tolerated temperatures of 16°F.

Bambusa multiplex var. riviereorum
Country of origin: China
Native name: Koon yam chuk

This is the dwarfest variety of the *multiplex* group, and is distinguished by the fact that the internodes are solid. The foliage is similar to cv. 'Fernleaf', and it is slightly hardier than the typical form. Known as the 'Chinese Goddess' bamboo.

Bambusa multiplex cv. 'Silverstripe'
Syn. *Bambusa argentea striata*
    Bambusa nana var. *argenteostriata*
    Bambusa nana var. *variegata*
    Bambusa vittata argentea
    Leleba multiplex f. *variegata*
Country of origin: China
Native name: Hosho-chiku

B. multiplex cv. 'Silverstripe' is believed to be the largest form of *multiplex*. The canes are somewhat similar in form to those of *multiplex*. but have a few threadlike white stripes. The cane sheaths are green with several brown lines, the brown colour remaining long after the green has faded. Can endure slightly lower temperatures than the typical form. Grown extensively in the southern and south-western states of the U.S.A., where it is known as the 'silverstripe hedge' bamboo.

Bambusa mutiplex cv. 'Silverstripe Fernleaf'
Syn. *Bambusa nana* f. *albo variegata*
  Leleba floribunda f. *albo variegata*
Country of origin: China
Native name: Fuiriho-o-chiku

This bamboo is similar to cv. 'Fernleaf', except that the leaves are white striped. It rather rare, and its hardiness is akin to cv. 'Silverstripe'.

Bambusa multiplex cv. 'Stripestem Fernleaf'
Syn. *Bambusa nana* f. *viridi striata*
  Leleba floribunda f. *viridi striata*
Country of origin: China
Native name: Beniho-o-chiku

Somewhat similar to cv. 'Fernleaf', with the exception of the canes, which are coloured pale reddish or yellow, with irregular green striping at first. Hardiness akin to cv. 'Fernleaf'.

Bambusa multiplex cv. 'Willowy'
Countries of origin: China; Japan
Native name not known

This bamboo has mid-green canes which grow up to 20 feet in height by $\frac{3}{4}$ inch in diameter. Cane sheaths are green, fading to a dull straw shade. The lowest internodes are solid, and the higher ones thick walled. Branches and twigs are very slender. Leaves narrow, roughly $4\frac{1}{2}$ inches in length by $\frac{3}{8}$ inch wide. The fully grown canes tend to cascade or droop excessively with the masses of fine foliage. Cold resistance roughly similar to the typical form.

Bambusa tootsik
Syn. *Sinobambusa tootsik*
Countries of origin: China; Japan
Native name: To-chiku

CANES Can be as high as 15 feet by up to $1\frac{1}{4}$ inches in diameter. Mid-green in colour.
NODES Prominent. From 12 to 18 inches apart.
BRANCHES Long and slender. Three to nine leaves per branch.

LEAVES Lanceolate. From 2 to 8 inches in length by $\frac{1}{2}$ to $\frac{3}{4}$ inch wide. Bluntly wedge-shaped at the base, and short but sharply tapered at the tip. Mid-green.

Bambusa vulgaris
Country of origin: Native to the East Indies, China, India, and
Japan
Native name: Taisan-chiku

B. vulgaris, which is perhaps the most widely grown bamboo in the Orient, can be cultivated in Western gardens provided that it can be given a relatively frost-free and wind-free environment. It may lose a proportion of its foliage during frosty weather, but the canes and rhizomes are seldom damaged beyond recovery.

DISTINGUISHING FEATURES The absence of visible tessellation in the leaves. The cane sheaths are characteristic of the Bambusa in that the apex is wide and the centre arches gently.

CANES 15 to 20 feet in height by $\frac{3}{4}$ to 1 inch in diameter in mild climates, but much larger in warmer conditions. Mid-green at first, ageing to dull yellow.

NEW SHOOTS Late May to June. Bright green.

CANE SHEATHS Not persistent except on late-emerging canes. Apex very wide with the central portion arching. Coated with fine stiff adpressed hairs at first, then smooth. Dull green in colour at first, fading to a drab buff shade.

LEAVES Up to 9 inches in length by $1\frac{1}{4}$ inches wide. Lack visible tessellation. Edged on both margins with fine bristles. Bluntly rounded at the base and short tapered to end in a fine point at the tip. Midrib very fine but palpable for three-quarters of its length. Four to seven pairs of secondary veins. Bright glossy green on the upper surface, duller matt green on the lower.

PROPAGATION Basal cane cuttings. Division of the clump. Rhizome cuttings.

ROOTSTOCK Though classed as running, it will only do so in very warm environments.

Table of Synonyms

As the great majority of hardy bamboo species have been renamed and reclassified over and over again by various botanists, it has not been an easy matter for a layman to decide which name was correct. I have therefore listed below, in alphabetical order, all the known synonyms of the various species opposite the name by which each is generally recognized, and by which it is listed in Chapter 6, 'Descriptive Classification of the Species, Varieties, and Cultivars'.

ABBREVIATIONS

A. Arundinaria
B. Bambusa
C. Chimonobambusa
Ch. Chusquea
I. Indocalamus
L. Leleba
N. Nipponocalamus
P. Phyllostachys
Pl. Pleioblastus
Ps. Pseudosasa
S. Sasa

Sam. Sasamorpha
Se. Sasaella
Semi. Semiarundinaria
Sh. Shibataea
Sina. Sinarundinaria
Sb. Sinobambusa
T. Tetragonacalamus
Th. Thamnocalamus
Y. Yadakeya
Yu. Yushiania

SYNONYMS	GENERALLY ACCEPTED NAME
alba marginata, B.	veitchii, S.
albo-marginata, S.	veitchii, S.
albo-striata, B.	simonii var. variegata, A.
alphonse karri, B.	multiplex cv. 'Alphonse Karr', B.
angulata, B.	quadrangularis, B.
angustifolia, B.	angustifolia, A.
argentea, B.	multiplex, B.
argentea striata, B.	multiplex cv. 'Silverstripe', B.
argenteostriata, A.	chino var. argenteostriata, A.

argenteostriata, B.	chino var. argenteostriata, A.
argenteostriata var. communis, A.	communis, A.
argenteostriata var. disticha, A.	disticha, A.
argenteostriatus, N.	chino var. argenteostriata, A.
argenteostriatus cv. 'Disticha', N.	disticha, A.
aristatus, Th.	aristata, A.
aurea, B.	aurea, P.
aurea, Sina.	aurea, P.
aureus, P.	aurea, P.
bambusoides, P.	quilioi, P.
bambusoides var. sulphureus, P.	sulphurea, P.
bambusoides cv. 'Allgold', P.	sulphurea, P.
bambusoides cv. 'Castillon', P.	castillonis, P.
bambusoides var. alba marginata, P.	veitchii, S.
bambusoides var. aurea, P.	aurea, P.
bambusoides var. boryana, P.	boryana, P.
borealis, B.	borealis, S.
breviglumis, Ch.	culeou, Ch.
boryana, B.	boryana, P.
boryanus, P.	boryana, P.
castilloni, P.	castillonis, P.
castillonis, B.	castillonis, P.
chino, B.	chino, A.
chino, N.	chino, A.
chino, Pl.	chino, A.
chino var. laydekeri, A.	laydekeri, A.
chino var. laydekeri, Pl.	laydekeri, A.
chrysantha, A.	chrysantha, S.
chrysantha, B.	chrysantha, S.
communis, Pl.	communis, A.
disticha, B.	disticha, A.
disticha, S.	disticha, A.
distichus, Pl.	disticha, A.
edulis, B.	pubescens, P.
edulis, P.	pubescens, P.
erecta, B.	hindsii, A.
falcata, B.	falcata, A.
falconeri, Th.	falconeri, A.
fastuosa, B.	fastuosa, A.
fastuosa, Semi.	fastuosa, A.

fastuosa var. yashadake, Semi.	fastuosa var. yashadake, A.
fauriei, P.	henonis, P.
flexuosus, P.	flexuosa, P.
floribunda, B.	falconeri, A.
floribunda, L.	multiplex cv. 'Fernleaf', B.
floribunda f. albo variegata, L.	multiplex cv. 'Silverstripe Fernleaf', B.
floribunda f. viridi striata, L.	multiplex cv. 'Stripestem Fernleaf', B.
fortunei var. aurea, B.	auricoma, A.
fortunei var. viridis, A.	humilis, A.
fortunei variegata, B.	fortunei, A.
fulvus, P.	fulva, P.
gauntlettii, S.	gauntlettii, A.
gigantea, A.	macrosperma, A.
gracilis, B.	falcata, A.
graminea, B.	graminea, A.
gramineus, Pl.	graminea, A.
henonis, B.	henonis, P.
heterocycla, P.	mitis, P.
hindsii, Pl.	hindsii, A.
hindsii, Th.	hindsii, A.
hindsii var. graminea, A.	graminea, A.
humilis, Pl.	humilis, A.
intermedia, Sb.	intermedia, A.
japonica, B.	japonica, A.
japonica, Ps.	japonica, A.
japonica, S.	japonica, A.
japonica, Y.	japonica, A.
kagamiana, Semi.	fastuosa var. kagamiana, A.
kinsianus, Pl.	kinsiana, A.
kokantsik, A.	marmorea, A.
kumasaca, B.	kumasasa, Sh.
kuma-saca, P.	kumasasa, Sh.
kumasasa, P.	kumasasa, Sh.
kurilensis, A.	kurilensis, S.
kurilensis, Ps.	kurilensis, S.
kurilensis, B.	kurilensis, S.
kurilensis var. cernua, S.	cernua, S.
kurilensis var. paniculata, A.	senanensis, S.

kurilensis var. uchidae, S.	uchidae, S.
linearis, Pl.	linearis, A.
macrosperma suffruticosa, A.	tecta, A.
macrosperma var. tecta, A.	tecta, A.
marliaci, P.	marliacea, P.
marmorea, B.	marmorea, A.
marmorea, C.	marmorea, A.
marmorea f. variegata, C.	marmorea cv. 'Variegata', A.
marmorea var. variegata, C.	marmorea cv. 'Variegata', A.
matsumurae, A.	marmorea, A.
maximowiczii, A.	chino, A.
maximowiczii. Pl.	chino, A.
mazeli, B.	quilioi, P.
metallica, B.	palmata, S.
metake, A.	japonica, A.
metake, B.	japonica, A.
mitis, B.	mitis, P.
multiplex, L.	multiplex, B.
multiplex f. alphonse karri, L.	multiplex cv. 'Alphonse Karr', B.
multiplex f. variegata, L.	multiplex cv. 'Silverstripe', B.
murielae, Sina.	murielae, A.
nagashima, B.	humilis, A.
nagashima, N.	humilis, A.
nagashima, Pl.	humilis, A.
nana, B.	multiplex, B.
nana f. albo variegata, B.	multiplex cv. 'Silverstripe Fernleaf', B.
nana f. viridi striata, B.	multiplex cv. 'Stripestem Fernleaf', B.
nana var. alphonse karri, B.	multiplex cv. 'Alphonse Karr', B.
nana var. argentea, B.	multiplex, B.
nana var. argenteostriata, B.	multiplex cv. 'Silverstripe', B.
nana var. disticha, B.	multiplex cv. 'Fernleaf', B.
nana var. gracilloina, B.	multiplex cv. 'Fernleaf', B.
nana var. normalis, B.	multiplex, B.
nana var. variegata, B.	multiplex cv. 'Silverstripe', B.
narihira, A.	fastuosa, A.
newmanii, B.	macrosperma, A.
niger, P.	nigra, P.
nigra, B.	nigra, P.

nigra, Sina.	nigra, P.
nigrer boryanus, P.	boryana, P.
nigra cv. 'Bory', P.	boryana, P.
nigra cv. 'Henon', P.	henonis, P.
nigra var. punctata, P.	punctata, P.
nigra var. henonis, P.	henonis, P.
nigra var. henonis, Sina.	henonis, P.
niitakayamensis, I.	niitakayamensis, A.
nittakayamensis, S.	niitakayamensis, A.
nittakayamensis, Yu.	niitakayamensis, A.
niitakayamensis var. microcarpa, I.	niitakayamensis var. microcarpa, A.
nipponica, B.	nipponica, S.
nitida, Sina.	nitida, A.
nobilis, A.	falconeri, A.
owatarii, A.	owatarii, S.
owatarii, Ps.	owatarii, S.
owatarii, Y.	owatarii, S.
palmata, A.	palmata, S.
palmata, B.	palmata, S.
paniculata, S.	senanensis, S.
puberula, B.	henonis, P.
puberula, P.	punctata, P.
puberula var. boryana, P.	boryana, P.
puberula var. nigra, P.	nigra, P.
pubescens, Sina.	pubescens, P.
pubescens, Pl.	pygmaea, A. (pubescent form).
pumila, B.	pumila, A.
pumila, S.	pumila, A.
pumilus, N.	pumila, A.
pumilus, Pl.	pumila, A.
punctatus, P.	punctata, P.
purpurascens, A.	borealis, S.
purpurascens, S.	borealis, S.
purpurascens var. borealis, S.	borealis, S.
pygmaea, S. (pubescent form)	pygmaea, A. (pubescent form)
pygmaea, S. (glabrous form)	pygmaea, A. (glabrous form)
pygmaea, B.	pygmaea, A. (pubescent form)
pygmoea, B.	pygmaea, A. (glabrous form)
pygmaeus, Pl.	pygmaea, A. (pubescent form)
quadrangularis, C.	quadrangularis, B.

quadrangularis, T.	quadrangularis, B.
quilioi, B.	quilioi, P.
ragamowski, A.	tessellata, S.
ragamowski, B.	tessellata, S.
ramosa, A.	ramosa, S.
ramosa, B.	ramosa, S.
ramosa, Se.	ramosa, S.
reticulata, P.	quilioi, P.
reticulata var. aurea, P.	aurea, P.
ruscifolia, P.	kumasasa, Sh.
ruscifolia, S.	kumasasa, Sh.
senanensis, B.	senanensis, S.
simonii, B.	simonii, A.
simonii, N.	simonii, A.
simonii cv. 'Silverstripe', A.	simonii var. variegata. A.
simonii, Pl.	simonii, A.
simonii var. chino, A.	chino, A.
spathiflorus, Th.	spathiflora, A.
spiculosa, Ps.	borealis, S.
spiculosa, S.	borealis, S.
sulphureus, P.	sulphurea, P.
sulphurea var. holochrysa, P.	sulphurea, P.
sulphurea var. viridis, P.	mitis, P.
tessellata, B.	tessellata, S.
tessellata, Sam.	tessellata, S.
tootsik, Sb.	tootsik, B.
uchidae, Ps.	uchidae, S.
vaginatus, Pl.	vaginata, A.
vaginatus, N.	vaginata, A.
variabilis var. pygmaea, A.	pygmaea, A. (glabrous form)
variegata, S.	variegata, A.
variegata var. pygmaea, S.	pygmaea, A. (pubescent form)
variegata var. viridis forma glabra, S.	pygmaea, A. (glabrous form)
variegatus, Pl.	fortunei, A.
variegatus forma glabra, Pl.	pygmaea, A. (glabrous form)
veitchii, A.	veitchii, S.
veitchii, B.	veitchii, S.
veitchii forma minor, S.	veitchii, var. nana, S.
viminalis, B.	kumasasa, Sh.
violascens, P.	violescens, P.

vilmorinii, B.	augustifolia, A.
virens, N.	virens, A.
virens, Pl.	virens, A.
viminalis, B.	kumasasa, Sh.
viridi glaucescens, B.	viridi glaucescens, P.
viridis, P.	mitis, P.
viridistriata, B.	simonii, A.
viridistriata, A.	auricoma, A.
viridistriatus, Pl.	auricoma, A.
viridistriatus vagans, Pl.	vagans, A.
vittata argentea, B.	multiplex cv. 'Silverstripe', B.
yashadake, Semi.	fastuosa var. yashadake, A.

The following have no known synonyms:

anceps, A.
hookeriana, A.
khasiana, A.
pantlingii, A.
racemosa, A.
tessellata, A.
castillonis var. inversus, P.
niger var. muchisasa, P.
viridi glaucescens var. linculi, P.
multiplex cv. 'Willowy', B.
multiplex var. riviereorum, B.
palmata var. nebulosa, S.

CHINESE AND JAPANESE NOMENCLATURE

Ao-nezasa	*A. virens*
Azumashine	*A. chino*
Azuma-nezasa	*A. chino*
Azumazasa	*S. ramosa*
Beniho-o-chiku	*B. multiplex* cv. 'Stripestem silver-leaf'
Bioji-chiku	*P. pubescens*
Biotan-chiku	*P. pubescens*
Bundodake	*Sh. kumasasa*
Bungozasa	*Sh. kumasasa*

Chimakizasa	*S. palmata*
Chishimazasa	*S. kurilensis*
Chigo-kan-chiku	*A. marmorea* cv. 'Variegata'
Fuiriho-o-chiku	*B. multiplex* cv. 'Silverstripe Fern-leaf'
Fushiodake-shino	*A. kinsiana*
Gomaizasa	*Sh. kumasasa*
Gokidake	*A. communis*
Gosan-chiku	*P. aurea*
Gyoyo-chiku	*A. linearis*
Ha-chiku	*P. henonis*
Hakonedake	*A. vaginata*
Hirouzasa	*A. humilis*
Ho-o-chiku	*B. multiplex* cv. 'Fernleaf'
Horai-chiku	*B. multiplex*
Hosho-chiku	*B. multiplex* cv. 'Silverstripe'
Hotei-chiku	*P. aurea*
Hyondake	*P. castillonis*
Ke-nezasa	*A. pygmaea* (pubescent form)
Ke-oroshima-chiku	*A. pygmaea* (pubescent form)
Mizakozasa	*S. nipponica*
Moso-chiku	*P. mitis*
Mosodake	*P. mitis*
Nagaba-nemagaridake	*S. uchidae*
Narihiradake	*A. fastuosa*
Nemagaridake	*S. senanensis*
Ne-zasa	*A. pygmaea* (glabrous form)
Nigadake	*P. quilioi*
Ogon-chiku	*P. sulphurea*
Okamesaza	*Sh. kumasasa*
Okinadake	*A. chino* var. *argenteostriata*
Okumyamazasa	*S. cernua*
Orishima-chiku	*A. disticha*
Owodake	*P. henonis*
Rikuchudake	*A. fastuosa* var. *kagamiana*
Rito-chiku	*P. pubescens*
Ryuku-chiku	*A. linearis*
Shibo-chiku	*P. marliacea*
Shimadake	*P. castillonis*
Shiro-chiku	*P. nigra*

Shiwa-chiku	*P. marliacea*
Shinegawadake	*A. chino*
Shikakudake	*B. quadrangularis*
Shiho-chiku	*B. quadrangularis*
Sudore-yoshi	*A. pumila*
Sui-chiku	*P. henonis*
Suisho-chiku	*P. henonis*
Suo-chiku	*B. multiplex* cv. 'Alphonse Karr'
Suzudake	*S. borealis*
Taibo-chiku	*P. aurea*
Tai-min-chiku	*A. graminea*
Taisan-chiku	*B. vulgaris*
Tan-chiku	*P. henonis*
Tao-chiku	*P. henonis*
To-chiku	*B. tootsik*
Toocu-chu	*B. quadrangularis*
Unmou-chiku	*P. boryana*
Yadake	*A. japonica*
Yajino	*P. quilioi*
Yakibazasa	*S. veitchii* var. *nana*
Yakushimadake	*S. owatarii*
Yashadake	*A. fastuosa* var. *yashadake*

Glossary

Names of genera and specific epithets are in italics; other terms defined
are printed in roman.

adpressed: pressed on, pressed against
aerial roots: roots that appear above ground-level
albino: whitish, pale, lacking chlorophyll
albo-striatus (-a, -um): marked with white lines
anceps: two-edged, two-headed, facing two ways; in the case of the
  bamboo of this name, of dubious origin
angulatus (-a, -um): angled instead of round
angustifolius (a, -um): with narrow leaves
argenteus (-a, -um): silvered, with a silver tint or lustre
argenteostriatus (-a, -um): with silvery lines or streaks
aristatus (-a, -um): having awns like an ear of barley
arundinaceous: rush- or reed-like, having a culm like that of a grass
aureus (-a, -um): of golden colour, golden yellow
auricle: an ear-like appendage; (in the cane sheath) a small prominence
  flanking the ligule
auricomus (-a, -um): golden haired

bamboo smut (*Ustilago shiraiana*); parasitic fungus which affects certain
  types of bamboo
Bambusa: a genus of bamboo; the name often applied in the past to
  species whose botanical classification was at the time obscure
basal roots: roots growing vertically from above soil-level
blade: see limbus
bloom (on cane): whitish or pale blue-white waxy deposit
borealis (-e): northern, of the north
breviglumis (-e): having small bracts like scales

caespitose: growing in close tufts
caryopsis: one-seeded indehiscent fruit with pericarp fused to the seed
  coat
cernuus (-a, -um): nodding or drooping

Cheshunt Compound: commercial fungicidal compound, a mixture of
  ammonium carbonate and copper sulphate
Chimonobambusa: one of the groups of species into which the Arundinaria have been divided, separated by some botanists as a distinct
  genus
chino: of China
Chinobambusa (Bamboo of China): one of the groups of species into which
  the Arundinaria have been divided, separated by some botanists as a
  distinct genus
chrysanthus (*-a, -um*): golden yellow
Chusquea: genus of bamboos mostly native to South America
clone: group of cultivated plants produced vegetatively from one
  original seedling or stock
communis (*-e*): common, widely distributed, growing in company
concretion: coalescence
culm (cane): jointed stem of grasses

deciduus (*-a, -um*): shedding its leaves annually
Dendrocalamus: genus of bamboos, mainly tropical in origin
dentate: toothed, with tooth-like notches, like teeth of a saw
desiccated: exhausted of all moisture, dried up
distichus (*-a, -um*): arranged in two rows

edulis (*-e*): edible, eatable
efflorescence: powdery crust covering a surface
embryonic: undeveloped, immature
epidermis: outermost layer of cells
erectus (*-a, -um*): upright, perpendicular

falcatus (*-a, um*): sickle-shaped
fastuosus (*-a, -um*): bountiful, stately
filamentosus (*-a, -um*): threadlike
flexuosus (*-a, -um*): bending alternately right and left, zigzag
floribundus (*-a, -um*): flowering abundantly
forma minor: small or dwarf form

giganteus (*-a, -um*): gigantic; usually higher than the type
glabrous: smooth, free from hairs
glands: secretory organs in leaves of certain semi-tender bamboos
glaucous: dull bluish green, typically with matt surface

gracilis (-*e*): slender, gracefully slight in form
gramineus (*a, -um*): grass-like

heterocyclus (*a, -um*): having unequal round joints
holochrysus (-*a, -um*): entirely yellow
humilis (-*e*): of low growth

Indocalamus (reed of India): one of the groups of species into which the Arundinaria have been divided, separated by some botanists as a distinct genus
inflorescence: collective flower of a plant
intermedius (-*a, -um*): intermediate, between two forms
intermediary veins: fine veins which run parallel to and between the secondary veins of leaves
internode: space between the nodes or joints
inversus (-*a, -um*): opposite to, turned over

japonicus (-*a, -um*): of Japan

keeled: with projecting ridge(s) with flattish areas each side
kurilensis (-*e*): of the Kurile Islands

lanceloate: lance-shaped
leptomorph: with running rootstock
lignify: to become woody
ligule: the rim-like projection at junction of cane sheath and sheath blade, on the inner side
limbus: limb; the blade or leaf-like appendage growing from the apex of the sheath
linear: long, narrow and of uniform breadth

macerate: make soft by soaking
macrospermus (-*a, -um*): bearing long seeds
marmoreus (-*a, -um*): marble-like
melanconicum culm: fungoid disease affecting certain bamboos
metallicus (-*a, -um*): with a metal-like lustre
microcarpus (-*a, -um*): bearing small fruit
midrib: main artery or vein of leaf
mitis (-*e*): gentle, defenceless, without thorns or spines
morphology: study of the forms of plants or animals
multiplex: manifold

nanus (*-a, -um*): dwarf
nebulosus (*-a, -um*): clouded
niger (*nigra, nigrum*): black
nipponicus (*-a, -um*): of Japan
Nipponocalamus (reed of Japan): one of the groups of species into which the Arundinaria have been divided, separated by some botanists as a distinct genus
nitidus (*-a, -um*): shining, smooth or clear
nobilis (*-e*): noble or stately
node: knot or joint
normalis (*-e*): according to type

ovary: organ of female reproductive system, containing ovule

pachymorph: forming dense clumps
palmatus (*-a, -um*): lobed; divided into five lobes like a hand
paniculatus (*-a, -um*): having tufts or panicles of flowers
photosynthesis: the conversion of carbon dioxide into carbohydrates in parts of plants containing chlorophyll, under the influence of light
Phyllostachys: genus of bamboos (from *phyllon,* a leaf, and *stachys,* a spike)
pith: spongy cellular tissue in stems and branches
Pleioblastus: one of the groups of species into which the Arundinaria have been divided, separated by some botanists as a distinct genus
Poaceae: the grass family
prophyllum (pl. prophylla): a sheath, usually two-keeled, at the first node of a branch. It lacks blade, auricles, and ligule
puberulus (*-a, -um*): somewhat downy
pubescens: covered with hair or down
pumilus (*-a, -um*): dwarfish, low or little
punctatus (*-a, -um*): marked with dots or spots
purpurascens: becoming purple
pygmaeus (*-a, -um*): dwarf, very small

quadrangularis (*-e*): four cornered or quadrate

racemosus (*a-, -um*): in the form of a raceme (an inflorescence with stalked flowers at intervals along a stem)
ramosus (*-a, -um*): having many branches
reticulatus (*-a, -um*): netted, resembling network

rhizome: prostrate subterranean root-like stem emitting roots on the lower side, and sending up leafy shoots on the upper surface; it is used as a storage depot for food

rootstock: part of plant normally below the soil surface

ruscifolius (-a, -um): resembling Ruscus or 'butcher's broom'

Sasa (Saza): genus of bamboos (probably from two Chinese words, Hsai-chu, or small bamboo)

Sasamorpha: one of the groups of species into which the Arundinaria have been divided, separated by some botanists as a distinct genus

scale: rudimentary or degenerate leaf

secondary veins: veins running from base of a leaf towards the tip on either side of the midrib, considerably more prominent than the intermediary veins

Semiarundinaria: one of the groups of species into which the Arundinaria have been divided, separated by some botanists as a distinct genus

septum (pl. septa): partition or dividing wall of tissue

seta (pl. setae): stiff hair or bristle

sheath: investing membrane, sheath-like covering

Shibataea: group of bamboos formerly belonging to the Arundinaria

siliceous: containing silica

Sino-: of China

Sinarundinaria, Sinobambusa: two of the groups of species into which the Arundinaria have been divided, separated by some botanists as distinct genera

spathiflorus (-a, -um): bearing flowers in spathes (i.e. one or a number of flowers enclosed in a bract)

spiculosus (-a, -um): bearing small spikes (as ears of corn)

stomata (pl. of stoma): minute orifices in epidermis of plants; mouth-like organs

striatus (-a, -um): scored on the surface; marked with fine longitudinal lines, grooves or flutings

suffruticosus (-a, -um): shrub-like

sulcus (pl. sulci): a groove

sulphureus (-a, -um): yellow or sulphur coloured

tabaschir (tabasheer, tabaxir): a siliceous concretion found in the diseased joint cavities of certain tropical bamboos

tectus (-a, -um): roof-like

tessellation: the mosaic of small unequal squares or oblongs formed by
  the veins of a leaf
Thamnocalamus (bushy reed): one of the groups of species into which
  the Arundinaria have been divided, separated by some botanists as a
  distinct genus
tillering: putting up shoots from the bottom of the original stalk

vagans: roving or wandering
vaginatus (-a, -um): sheath-like
variegation: diversity of colour
venation: arrangement of veins in a leaf
verticillate: arranged in a whorl, as round a stem
violescens: becoming violet, violet coloured
virens: of a green colour
viridi-glaucescens: of a bluish-green colour
viridistriatus (-a, -um): marked or scored with green
viridus (-e): green
vittatus (-a, -um): striped longitudinally

witterings: small, often weak shoots growing at the base of a tree or
  shrub

List of Nurseries Supplying Bamboos

Many nurseries in various parts of the United Kingdom propagate hardy bamboos for sale to the general public, so for the benefit of those who wish to purchase plants I have compiled a list of recommended suppliers.

The list does not claim to be comprehensive, but those nurseries listed are recommended by the author, who knows from past experience that they stock only high-quality material covering a useful range of bamboo species.

HILLIER & SONS, Winchester, Hants.

TRESEDERS' NURERIES (TRURO) LTD., The Nurseries, Truro, Cornwall.

GEO. JACKMAN & SON, Woking Nurseries Ltd., Woking, Surrey.

MICHAEL HAWORTH-BOOTH, Farall Nurseries, Roundhurst, Nr. Haslemere, Surrey.

V. N. GAUNTLETT & CO. LTD., The Nurseries, Chiddingfold, Surrey.

BRITISH BAMBOO CANE COMPANY LTD., Lanivet, Bodmin, Cornwall.

G. REUTHE LTD., Fox Hill Hardy Plant Nurseries, Keston, Kent.

Bibliography

BEAN, W. J., *Trees and Shrubs Hardy in the British Isles* (6th ed., Murray, London, 1936).

CAMUS, E. G., *Les Bambusees* (Paul Lechavier, Paris, 1913).

CASEY, J. P., *Pulp and Paper*, Vol. I, 2nd ed. (Interscience Publishers, London, 1960).

CHANG, F. C. LINGMAN, *Sci. Jour.*, XVII, 617, 1938.

COHN, F. J., *Memoire Ueber Tabaschir* (late nineteenth century).

FREEMAN-MITFORD, A. B., *The Bamboo Garden* (Macmillan, London, 1896).

FREY-WYSSLING, A., *Ber. Deutsch. Botan. Ges.*, 48, 179, 1930.

HARTWRIGHT, T. U., *Planting Trees and Shrubs in Gravel Workings* (Sand & Gravel Assn. of G.B., London, 1960).

HAUN, J. R., and YOUNG, R. A., *Bamboo in the United States*, U.S. Min. of Ag. Handbook, No. 193, 1961.

HAYATA, B., *Report on A. niitakayamensis, Bot. Mag.*, Tokyo, Vol. 21, p. 49, 1907.

KING, E. J., and BELT, T. W., *The Physiological and Pathological Aspects of Silica, Physio. Rev.*, Vol. 18, No. 3, 1938.

MACIE, J. L., *Phil. Trans.*, LXXXI, 368, 1791.

McCLURE, F. A., *The Bamboos, a fresh perspective* (Harvard University Press, 1966).

MELLOR, F. W., *Comprehensive Treatise of Inorganic and Theoretical Chemistry* (Longmans, Green, London, VI, 141, 1940).

MUNRO, WILLIAM, 'A monograph on the Bambusaceae', Report from *Linnean Society of London Trans.*, Vol. XXVI, Pt. I, London, 1868.

OHWI, 'The Illustrated Flora of Japan', *Trans. Smithsonian Inst.*, 1966.

REHDER, A., *Manual of Cultivated Trees and Shrubs Hardy in North America* (Macmillan, 2nd ed., 1940).

R.H.S. *Dictionary of Gardening* (Oxford University Press, 1951).

ROSEWARNE EXPERIMENTAL HORTICULTURE STATION, Camborne, Cornwall, *Shelter Trees and Shrubs*, Leaflet No. 2, 1963.

ROXBURGH, W., *Plants of the Coronel Coast* (Nicol, London).

Bibliography

Shepherd, F. W., *The Propagation of Arundinaria japonica*, Min. of Ag. Experimental Horticulture, No. 5, H.M.S.O., 1961.
Sokolov, S. Y., *The Trees and Shrubs of the U.S.S.R.*, Vol. 2, 1951.
United States Dispensatory (Lippincott, 23rd ed., 1943, 1516).
Venkatraman, T. S., *Indian Jour. Agr. Science*, 513, VII, 1937.

Index

A number in bold type against the generally accepted name of a species or variety
indicates the page on which the detailed description appears.